Simon Kuhn

Transport Mechanisms in Mixed Convective Flow over Complex Surfaces

Simon Kuhn

Transport Mechanisms in Mixed Convective Flow over Complex Surfaces

An experimental and numerical study

Südwestdeutscher Verlag für Hochschulschriften

Impressum/Imprint (nur für Deutschland/ only for Germany)
Bibliografische Information der Deutschen Nationalbibliothek: Die Deutsche Nationalbibliothek verzeichnet diese Publikation in der Deutschen Nationalbibliografie; detaillierte bibliografische Daten sind im Internet über http://dnb.d-nb.de abrufbar.

Alle in diesem Buch genannten Marken und Produktnamen unterliegen warenzeichen-, marken- oder patentrechtlichem Schutz bzw. sind Warenzeichen oder eingetragene Warenzeichen der jeweiligen Inhaber. Die Wiedergabe von Marken, Produktnamen, Gebrauchsnamen, Handelsnamen, Warenbezeichnungen u.s.w. in diesem Werk berechtigt auch ohne besondere Kennzeichnung nicht zu der Annahme, dass solche Namen im Sinne der Warenzeichen- und Markenschutzgesetzgebung als frei zu betrachten wären und daher von jedermann benutzt werden dürften.

Verlag: Südwestdeutscher Verlag für Hochschulschriften Aktiengesellschaft & Co. KG
Dudweiler Landstr. 99, 66123 Saarbrücken, Deutschland
Telefon +49 681 37 20 271-1, Telefax +49 681 37 20 271-0, Email: info@svh-verlag.de
Zugl.: Zürich, ETH, Diss., 2008

Herstellung in Deutschland:
Schaltungsdienst Lange o.H.G., Berlin
Books on Demand GmbH, Norderstedt
Reha GmbH, Saarbrücken
Amazon Distribution GmbH, Leipzig
ISBN: 978-3-8381-0759-2

Imprint (only for USA, GB)
Bibliographic information published by the Deutsche Nationalbibliothek: The Deutsche Nationalbibliothek lists this publication in the Deutsche Nationalbibliografie; detailed bibliographic data are available in the Internet at http://dnb.d-nb.de.

Any brand names and product names mentioned in this book are subject to trademark, brand or patent protection and are trademarks or registered trademarks of their respective holders. The use of brand names, product names, common names, trade names, product descriptions etc. even without a particular marking in this works is in no way to be construed to mean that such names may be regarded as unrestricted in respect of trademark and brand protection legislation and could thus be used by anyone.

Publisher:
Südwestdeutscher Verlag für Hochschulschriften Aktiengesellschaft & Co. KG
Dudweiler Landstr. 99, 66123 Saarbrücken, Germany
Phone +49 681 37 20 271-1, Fax +49 681 37 20 271-0, Email: info@svh-verlag.de

Copyright © 2009 by the author and Südwestdeutscher Verlag für Hochschulschriften Aktiengesellschaft & Co. KG and licensors
All rights reserved. Saarbrücken 2009

Printed in the U.S.A.
Printed in the U.K. by (see last page)
ISBN: 978-3-8381-0759-2

Abstract

In this thesis transport mechanisms in forced and mixed convective flows over complex surfaces are addressed by experiments and numerical simulations. Compared to classical channel flow the surface structure adds a degree of complexity to the flow by inducing streamline curvature, flow separation and flow reattachment, thus leading to flow situations which are often present in relevant technical and geophysical applications. In order to achieve homogeneous and inhomogeneous reference flow situations two different well–defined types of bottom surfaces are considered: (i) three two–dimensional wavy walls with amplitude–to–wavelength ratios of $\alpha = 2a/\Lambda = 0.1$ ($\Lambda = 30$ mm), $\alpha = 0.2$ ($\Lambda = 30$ mm), and $\alpha = 0.2$ ($\Lambda = 15$ mm), and (ii) a three–dimensional profile consisting of two superimposed sinusoidal waves.

Measurements are carried out in a wide water channel facility (aspect ratio 12:1) where the bottom surface is heated with an incorporated electrical heat foil. Digital particle image velocimetry (PIV) is performed to examine the spatial variation of the streamwise, spanwise and wall-normal velocity components in three different measurement planes. Planar laser induced fluorescence (LIF) is used to measure the concentration of a species emanating from a point source, or to determine the temperature field in the heated cases. In case of forced convection measurements are performed at Reynolds numbers ranging from 5600 to 40000, defined with the channel height H and the bulk velocity U_B. For the mixed convection, which is obtained by applying a large temperature gradient between the top and bottom wall, the measurements are conducted at Reynolds numbers of $Re_H = 1100$ and Grashof numbers of $Gr_H = 1.94 \cdot 10^6$, respectively $Re_H = 2120$ and $Gr_H = 1.30 \cdot 10^6$. The numerical simulations are performed employing a large eddy simulation (LES) with a dynamic subgrid scale model.

The resulting momentum fields and scalar fields are characterized by calculating turbulence statistics and the scalar fluxes. Structural information is obtained by performing a proper orthogonal decomposition (POD) of the velocity components to extract the most energetic flow structures, and additionally by calculating the swirling strength.

Our results show a good agreement between experiment and numerical simulation. In the case of forced convection the transport of a scalar species is enhanced by the presence of the wavy surfaces. Comparing the influence of the different types of surfaces the superimposed waves promote the spreading of the scalar plume most. These enhanced transport properties are attributed to large–scale structures present in the flow field. These coherent structures exhibit a dependency in their spatial organization and their energy contribution on the wavy surfaces. Advancing to mixed convective flows the transport of momentum and scalars is additionally enhanced through buoyancy induced fluid motions. This leads to an augmentation of the transport of heat away from the heated wavy surface, which is expressed by increased Nusselt numbers compared to theoretical predictions. The patterns of the Nusselt number on the heated surface and the magnitude of the heat flux exhibit a correlation on the spatial distribution of coherent structures. Characterizing the state of isotropy of the flow it is found that the turbulence state in the center of the flow channel is more anisotropic than in classical channel flow, resulting in an increased transport normal to the mean flow direction.

The major finding is the presence of longitudinal, coherent flow structures which are found for each flow case and which are directly linked to the transport mechanisms for momentum and scalars. These results suggest that 'controlling' these large–scale structures provides means to directly influence transport properties in technical applications.

Zusammenfassung

Die vorliegende Arbeit befasst sich mit Transportmechanismen in Zwangs- und Mischkonvektion über komplexen Oberflächen und untersucht diese mittels experimenteller und numerischen Methoden. Im Vergleich zur klassischen Kanalströmung erhöht die Struktur der Oberflächen die Komplexität des betrachteten Systems: die Stromlinien werden gekrümmt, es kommt zu Strömungsablösung und Wiederanlagerung. Diese Phänomene zeichnen viele in der Technik und Geophysik vorkommende Strömungen aus. Um sowohl homogene als auch inhomogene Strömungszustände zu erreichen werden folgende Arten von Bodengeometrien betrachtet: (i) drei zwei-dimensionale gewellte Wände mit einem Amplituden-zu-Wellenlängenverhältnis von $\alpha = 2a/\Lambda = 0.1$ ($\Lambda = 30$ mm), $\alpha = 0.2$ ($\Lambda = 30$ mm), und $\alpha = 0.2$ ($\Lambda = 15$ mm), sowie (ii) eine drei-dimensionale Oberflächenstruktur die durch die orthogonale Überlagerung zweier Sinuswellen gebildet wird.

Die Experimente wurden in einem Strömungskanal mit einem Seitenverhältnis (Breite zu Höhe) von 12:1 durchgeführt, dessen Bodenwand durch eine eingebettete Heizfolie beheizt wurde. Die räumliche Änderung der Geschwindigkeitskomponenten in allen Raumrichtungen wurde mittels digitaler Particle-Image-Velocimetry (PIV) in drei Messebenen bestimmt. Mit der Methode der laserinduzierten Fluoreszenz (LIF) wurde die Konzentrationsverteilung einer Spezies, die dem Strömungskanal über eine Punktquelle zugeführt wurde, ermittelt, bzw. im Falle der Mischkonvektion das Temperaturfeld bestimmt. Die Messungen in Gebiet der Zwangskonvektion umspannten einen Bereich der Reynoldszahl (bestimmt mit der Kanalhöhe und der mittleren Strömungsgeschwindigkeit) von 5600 bis 40000. Im Falle der Mischkonvektion wurden die Experimente für die Reynoldszahl $Re_H = 1100$ und die Grashofzahl $Gr_H = 1.94 \cdot 10^6$, bzw. $Re_H = 2120$ und $Gr_H = 1.30 \cdot 10^6$, durchgeführt. Der Strömungszustand der Mischkonvektion wurde durch das Aufbringen eines Temperaturgradienten zwischen Boden- und Deckelplatte des Kanals erreicht. Die Simulationen wurden mit der Methode der Grobstruktursimulation (LES) mit dynamischen Modell zur Bestimmung der Feinstrukturanteile durchgeführt.

Die resultierenden Impuls- und Skalarfelder werden durch die Bestimmung von Turbulenzgrössen und skalaren Transporttermen charakterisiert. Eine Eigenwertzerlegung (POD; proper orthogonal decomposition) wird verwendet um strukturelle Information über die energiereichsten Strukturen der Strömung zu gewinnen, desweiteren wird zur zusätzlichen Charakterisierung die Wirbelstärke (swirling strength) berechnet.

Die erhaltenen Resultate zeigen eine gute Übereinstimmung zwischen Experiment und Simulation. Im Falle der Zwangskonvektion wird der Skalartransport durch die Anwesenheit der komplexen Oberflächen verstärkt. Vergleichend wird festgestellt, dass die drei-dimensionale Oberflächenstruktur die Aufweitung der Rauchfahne der skalaren Spezies am meisten erhöht. Diese Erhöhung des Skalartransports wird auf die Gegenwart von Grobstrukturen im Strömungsfeld zurückgeführt. Deren räumliche Anordnung und die Energieverteilung dieser kohärenten Strukturen weist eine Abhängigkeit von der überströmten Oberfläche auf. Im Bereich der Mischkonvektion werden die Transportvorgänge zusätzlich durch Fluidbewegungen, hervorgerufen durch die lokal verringerte Dichte, verstärkt. Dies führt zu einer Erhöhung des Wärmetransports von der geheizten Wand auf das Fluid, dies spiegelt sich in höheren Werten für die Nusseltzahl im Vergleich zu theoretischen Vorhersagen wieder. Die räumliche Verteilung der Nusseltzahl sowie das Maximum des Wärmetransports ist korreliert mit der räumlichen Organisation der Grobstrukturen. Die Bestimmung der Anisotropie der Strömung im Bereich der Mischkonvektion zeigt, dass diese in der Kanalmitte im Vergleich zur klassischen Kanalströmung erhöht ist. Hieraus resultiert ein erhöhter Transport von Impuls senkrecht zur Hauptströmungsrichtung.

Eines der wichtigsten Ergebnisse dieser Untersuchungen ist das Auffinden von strömungsorientierten Grobstrukturen, die einen direkten Einfluss auf den Impuls- und Skalartransport haben. Diese Resultate deuten auf die Möglichkeit hin, dass eine 'Kontrolle' dieser Grobstrukturen eine direkte Beeinflussung von Transportvorgängen in technischen Anwendungen erlaubt.

Contents

Abstract   1

Zusammenfassung   3

Nomenclature   9

1 Introduction   1

2 Theoretical Considerations   5
    2.1 Governing Equations for Turbulent Flows 5
        2.1.1 Reynolds Stress Equation 6
        2.1.2 Turbulent Heat Flux Equation 7
    2.2 Scales of Turbulent Motion 7
    2.3 Wall Influence on Turbulence 8
        2.3.1 Flat Wall 8
        2.3.2 Wavy Wall 9
    2.4 Proper Orthogonal Decomposition 13
        2.4.1 Mathematical Background 13
        2.4.2 Method of Snapshots 14

3 Experiments and Experimental Techniques   17
    3.1 Channel Facility 17
        3.1.1 Test Section 19
        3.1.2 Experiments 20
    3.2 Particle Image Velocimetry 22
        3.2.1 Geometric Imaging 23
        3.2.2 Tracer Particles 24
        3.2.3 Measurement Accuracy 26
        3.2.4 Spatial Resolution 28
    3.3 Laser Induced Fluorescence 30
        3.3.1 One–color Laser Induced Fluorescence 30
        3.3.2 Two–color Laser Induced Fluorescence 31
        3.3.3 Properties of Tracer Dyes 31
        3.3.4 Measurement Accuracy 33
    3.4 Simultaneous Velocity and Scalar Measurements 34
        3.4.1 Simultaneous Velocity and Concentration Measurements 34
        3.4.2 Simultaneous Velocity and Temperature Measurements 35

4 Influence of Wavy Surfaces on Coherent Structures in a Turbulent Flow   39
    4.1 Introduction 39

4.2 Results 40
    4.2.1 Results in the (x,y)–plane 40
    4.2.2 Results in the (x,z)–plane 42
    4.2.3 Results in the (y,z)–plane 47
4.3 Conclusions 48

5 The Influence of Wavy Walls on the Transport of a Passive Scalar in Turbulent Flows   51
5.1 Introduction 51
5.2 Flow Field 53
    5.2.1 Mean Velocity 53
    5.2.2 Shear Layer 55
5.3 Scalar Transport 55
    5.3.1 Concentration Field 55
    5.3.2 Spreading of the Scalar Plume 58
    5.3.3 Scalar Fluxes 63
5.4 Conclusions 64

6 Experimental Study of Heat Flux in Mixed Convective Flow over Solid Waves   69
6.1 Introduction 69
6.2 Flow Description 71
6.3 Results 72
    6.3.1 Mean Velocity Profiles 72
    6.3.2 Root Mean Square of Velocity Fluctuations 73
    6.3.3 Reynolds Stress 74
    6.3.4 Mean Temperature Field and Scalar Fluxes 75
    6.3.5 Proper Orthogonal Decomposition of the Velocity Field 76
    6.3.6 Scalar Fluxes 76
6.4 Conclusions 80

7 The Influence of Mixed Convection on the Transport Properties of a Scalar Species   83
7.1 Introduction 83
7.2 Results 85
    7.2.1 Velocity Field 85
    7.2.2 Scalar Field 88
7.3 Conclusions 95

8 Numerical Simulation of Mixed Convection over a Wavy Wall: A Dynamical LES Approach   101
8.1 Introduction 101
8.2 Flow Description 103
8.3 Subgrid Models and Numerical Details 105
8.4 Flow Field 107
    8.4.1 Velocity Profiles 107
    8.4.2 Turbulent Kinetic Energy, Turbulence Anisotropy and Coherent Structures 112
8.5 Scalar Field 116
    8.5.1 Mean Temperature and Temperature Variance 116
    8.5.2 Turbulent Heat Flux and Wall Heat Transfer 123

8.6 Conclusions	131
9 Conclusions	**133**
10 Outlook	**135**
10.1 Reactive Flows	135
10.2 Measurement Techniques	135
10.2.1 Tomographic PIV	135
10.2.2 Multiphase flows	136
10.3 Hybrid Modeling Techniques	136
A Principles of Large Eddy Simulations	**137**
A.1 Filtering Operations in LES	137
B Optical Devices for Two-Color Laser Induced Fluorescence	**139**
B.1 Beamsplitter	139
B.2 Filter to Remove Mie-Scattering	140
Bibliography	**141**
Acknowledgements	**149**

Nomenclature

Roman Symbols

a	half–amplitude of the wave profile	[m]
	constant in the Basset equation	
a_{ij}	anisotropy tensor	
A	anisotropy parameter	
b	constant in the Basset equation	
B	channel width	[m]
c	concentration	[mol/l]
	constant in the Basset equation	
c_f	friction factor	
c_p	specific heat capacity at constant pressure	[J/(kgK)]
C_S	Smagorinsky constant	
$\mathbf{C}$	covariance matrix	
d_e	particle image diameter	[m]
D	diffusion coefficient	[m^2/s]
d_P	particle diameter	[m]
D_a	aperture diameter	[m]
f	focal length	[m]
$f_\#$	f–number of the camera objective	
F_i	external force	[N]
g	acceleration due to gravity	[m/s^2]
Gr_H	Grashof number for channel flow, $gH^3\Delta T\beta/\nu^2$	
h	half channel height	[m]
H	full channel height	[m]
k	turbulent kinetic energy	[m^2/s^2]
l	lengthscale	[m]
l_0	lengthscale of the largest eddies	[m]
l_{DI}	demarcation lengthscale between the dissipation range ($l < l_{DI}$) and the inertial subrange ($l > l_{DI}$)	[m]
l_{EI}	demarcation lengthscale between the energy–containing range of eddies ($l > l_{EI}$) and smaller eddies ($l < l_{EI}$)	[m]
l_r	smallest resolvable lengthscale by PIV	[m]
$\mathcal{L}$	characteristic lengthscale of the flow	[m]
M	magnification factor	
	number of velocity fields (ensemble size)	
N_S	Stokes number	
Nu	Nusselt number	
p	pressure	[Pa]
Pr	Prandtl number	

$\Pr_t$	turbulent Prandtl number	
$\dot{q}$	heat flux	$[\text{W}/\text{m}^2]$
R	two–point correlation tensor	
Ra_H	Rayleigh number for channel flow, $\text{Gr}_H \Pr$	
Re_H	Reynolds number for channel flow, $U_B H/\nu$	
Ri	Richardson number, $\text{Gr}_H/\text{Re}_H^2$	
S	spreading rate	
Sc	Schmidt number, ν/D	
s_{ij}	rate–of–strain tensor	
t	time coordinate	$[\text{s}]$
T	temperature	$[\text{K}]$
$\mathbf{u}$	velocity vector (u,v,w)	
u	streamwise velocity component	$[\text{m/s}]$
$\overline{u}$	mean streamwise velocity component	$[\text{m/s}]$
u'	fluctuating streamwise velocity component	$[\text{m/s}]$
u_P	particle velocity	$[\text{m/s}]$
u_η	Kolmogorov velocity scale	$[\text{m/s}]$
u_w^*	friction velocity	$[\text{m/s}]$
U_B	bulk velocity	$[\text{m/s}]$
v	wall–normal velocity component	$[\text{m/s}]$
$\overline{v}$	mean wall–normal velocity component	$[\text{m/s}]$
v'	fluctuating wall–normal velocity component	$[\text{m/s}]$
w	spanwise velocity component	$[\text{m/s}]$
$\overline{w}$	mean spanwise velocity component	$[\text{m/s}]$
w'	fluctuating spanwise velocity component	$[\text{m/s}]$
x	streamwise coordinate direction	
x_R	reattachment point	
x_S	separation point	
$\mathbf{X}$	set of spatiotemporal data	
y	wall–normal coordinate direction	
y_w	wall profile	
z	spanwise coordinate direction	
$z_{1/2}$	half width of the scalar plume	$[\text{m}]$

Greek Symbols

α	amplitude to wavelength ratio	
	diffusor angle	$[°]$
β	angle between the (y_β,z)–plane and the x–axis	$[°]$
	phase angle in the Basset equation	$[\text{rad}]$
	volumetric thermal expansion coefficient	$[\text{m}^3/\text{kg}]$
δ	boundary layer thickness	
δ_z	depth of field	$[\text{m}]$
δ_{ij}	Kronecker symbol	
Δ	LES filter width	

ε	viscous dissipation term	
η	Kolmogorov lengthscale	[m]
	amplitude ratio in the Basset equation	
ϑ	temperature	[°C]
λ	light wavelength	[m]
	thermal conductivity	[W/(mK)]
λ_i	eigenvalue of POD mode i	
Λ	wavelength of the sinusoidal wall	[m]
Λ_z	spanwise lengthscale	[m]
μ	dynamic viscosity	[Pa s]
ν	kinematic viscosity	[m²/s]
ν_t	turbulent viscosity	[m²/s]
$\xi(\mathbf{x},t)$	scalar field of spatiotemporal data	
Π_i	eigenfunction of POD mode i	
ϱ	fluid density	[kg/m³]
ϱ_P	particle density	[kg/m³]
τ	shear stress	[Pa]
τ_w	wall shear stress	[Pa]
τ_η	Kolmogorov time scale	[s]
τ_{ij}	second–order stress tensor	
ϕ_i	eigenvector	
ω	circular frequency of fluid/particle motion	[1/s]

Abbreviations

AOV	area of view	[m²]
CL–2	Craik-Leibovich type 2 instability	
CCD	charge coupled device	
DES	detached eddy simulation	
DNS	direct numerical simulation	
DPIV	digital particle image velocimetry	
LES	large eddy simulation	
LIF	laser induced fluorescence	
Nd:YAG	neodymium:yttrium aluminium garnet ($Y_3Al_5O_{12}$) crystal	
PIV	particle image velocimetry	
PLIF	planar laser induced fluorescence	
POD	proper orthogonal decomposition	
PVC	poly vinyl chloride	
RANS	Reynolds averaged Navier–Stokes equations	
rms	root mean square	
1D	one–dimensional	
2D	two–dimensional	
3D	three–dimensional	
$\mathcal{O}\{\cdot\}$	order of $(\cdot)$	
$\overline{(\cdot)}$	time average	

$(\cdot)'$	fluctuation
$\langle \cdot \rangle$	spatial average
$\langle \cdot, \cdot \rangle$	Euclidean inner product

Subscript and Superscript

B	bulk quantity
H	channel height (used as length scale)
i	running index
j	running index
k	running index
rms	root-mean-square
w	wall quantity
$+$	wall units

Chapter 1
Introduction

In this thesis forced convective and mixed convective flows over complex surfaces are addressed. Thus we make an effort to investigate and describe relevant flow situations. Flows occurring in most technical applications and in nature (e.g. the atmospheric boundary layer) are characterized by high Reynolds number, complex bounding surfaces, transport of passive and active scalars, and the presence of multiple phases. The use of wavy walls as bounding surfaces reproduces the wall complexity in a mathematically well described manner. Throughout this thesis the forced and mixed convective single phase channel flow between a flat top wall and two different types of wavy bottom surfaces is investigated. The different complex surfaces are three two–dimensional wavy walls propagating in streamwise direction, and a three–dimensional surface formed by the superposition of two sinusoidal waves. In the following we shortly review the range of applications of this particular flow situation.

Technical applications Wavy walls are often incorporated in heat exchangers to enhance the heat transfer from the wall to the fluid (e.g. Rush et al. (1999); Metwally and Manglik (2004)). This heat transfer enhancement is attributed to the formation of longitudinal flow structures, formed by the interaction of the flow with the wavy surfaces.

Geophysical flow situations Two– and three–dimensional wavy surfaces are often used as model for complex geometrical situations occurring in nature, such as the turbulent flow over hills and waves (e.g. Belcher and Hunt (1998)). Recently numerical studies addressed sediment transport in turbulent flows over ripples (e.g. Zedler and Street (2001); Chang and Scotti (2003)). In these numerical works the surface complexity is expressed in form of two–dimensional sinusoidal surfaces, respectively the three–dimensional superposition of those. The complex boundaries in these studies are characterized by an amplitude–to–wavelength ratio α of 0.1, which is identical to the geometrical scaling of the surfaces in our laboratory studies. Another important aspect is the effect of surface complexity on relevant transport processes occurring in the atmospheric boundary layer. Raupach and Finnigan (1997) investigated the influence of topography on meteorological variables and surface–atmosphere interactions, Katul et al. (2006) addressed the influence of hilly terrain on canopy–atmosphere carbon dioxide exchange.

Wall influence on turbulence Wall–bounded flows have been an active area in turbulence research for decades, and the discussion about many aspects of this flow situation is still ongoing. As an example might serve that even topics already incorporated in many textbooks like the universality of the log–law for turbulent wall–bounded flows have been questioned recently (George (2007)). Especially when advancing from smooth surfaces to rough and undulated surfaces the question of how coherent structures are formed and sustained is of importance (e.g. Robinson (1991); Jiménez (2004)). The bulk of the momentum and scalar transport is associated with these large–scale, coherent structures.

Mixed convective flows   Mixed convective flows, i.e. flows where the transport of momentum and scalars is involved as a combination of free and forced convection, are present in technical applications, such as the aforementioned heat transfer devices, as well as in geophysical flow situations. An example for geophysical flows over complex boundaries is given in the work of Banna et al. (2004), in which the influence of buoyancy effects on flow structures and scalar transport processes in a two–dimensional vegetation canopy is addressed. Results show an enhancement of transport processes due to the formation of secondary flow structures in the flow field induced by buoyancy driven instabilities.

In previous studies the large–scale structures in a turbulent flow over a wavy wall for Reynolds numbers Re_H in a range between 100 and 14600 were investigated (Günther (2001a); Günther and Rudolf von Rohr (2003)). A wavy wall characterized by the amplitude–to–wavelength ratio of $\alpha = 0.1$ (Λ=30 mm) was used. Structural information was obtained by the method of proper orthogonal decomposition (POD) and a characteristic spanwise scale Λ_z in the most dominant modes of $\mathcal{O}\{1.5H\}$, where H denotes the channel height, was identified. This characteristic spanwise scale was confirmed to be almost identical for the first two eigenfunctions of the decomposed flow field from measurements over three different two–dimensional wavy surfaces (Kruse et al. (2006)). The turbulence quantities and large–scale structures were found to be independent of the Reynolds number in the outer region of the flow, when scaled with the outer velocity scale, U_B (Kruse (2005)). In addition, Kruse and Rudolf von Rohr (2006) investigated the transport of heat (as a passive scalar) in a turbulent flow over a heated wavy wall. By employing a particle image thermometry technique the velocity and temperature fields were measured simultaneously. Quantitative agreement between large–scale thermal and momentum structures was found. In a numerical study, Wagner (2007) addressed transport properties of non–isothermal, forced convective flows over two–, and three–dimensional wavy walls by means of large eddy simulation, detached eddy simulation and different turbulence models.

The present work advances these studies by addressing buoyancy effects induced by mixed convection from a wavy surface on transport processes by applying a combined digital particle image velocimetry and laser induced fluorescence technique to simultaneously measure the velocity, the temperature field, and the concentration field of a tracer species. The main aspects of these investigated flow situations can be summarized as follows:

- Flow separation and reattachment due to the interaction of the flow with the complex bounding surfaces

- Characteristic regions of maximum positive and negative shear stress

- High Reynolds numbers for isothermal flow, buoyancy effects in non–isothermal flow conditions

- Deformation of the mean velocity profile due to buoyancy induced motions

- Transport of a species introduced by a point source as passive scalar

- Transport of heat as an active scalar

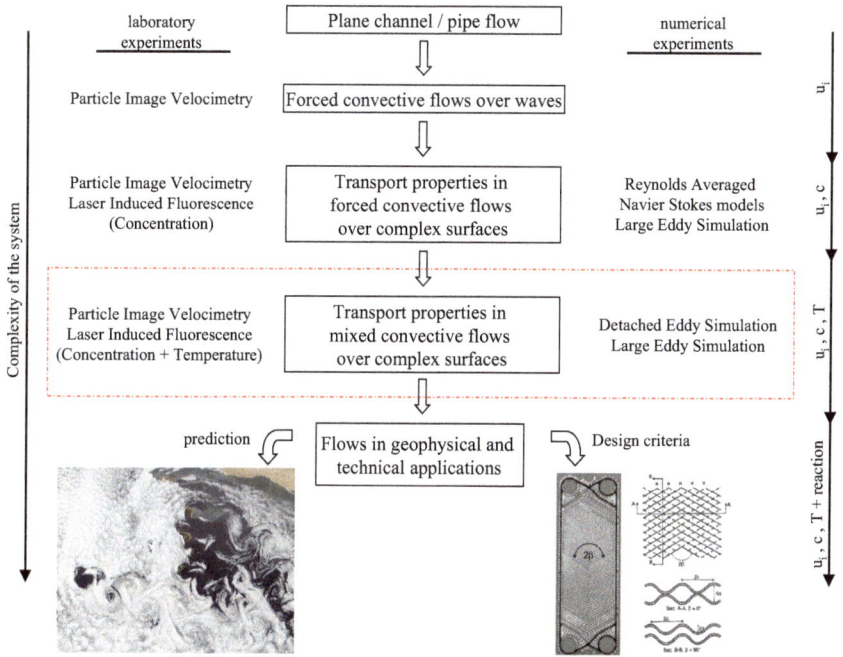

Figure 1.1: Outline and classification of the present project.

The outline and classification of this project are depicted in Figure 1.1.

This thesis is organized as follows: In CHAPTER 2 a description of the fundamental aspects of turbulent flows and of wall influence on turbulence and coherent structures is provided. The governing momentum and scalar equations, the different scales present in turbulent flows, and the influence of flat and complex walls on turbulent flows and on coherent structures are addressed. The chapter is concluded with the fundamentals of the proper orthogonal decomposition (POD) which is used as a tool to extract coherent structures. CHAPTER 3 presents the experimental facility and the applied experimental techniques. To measure the velocity field particle image velocimetry (PIV) is used, the scalar fields are assessed by means of laser induced fluorescence (LIF). Furthermore, the experimental setup for simultaneous measurements of both fields is described. In CHAPTER 4 we address the influence of different complex surfaces on coherent structures. A difference in the spanwise scale of the dominant eigenmode is found by comparing sinusoidal profiles with a three–dimensional surface. The influence of these coherent structures on turbulent scalar transport is discussed in CHAPTER 5, where the spreading of a scalar plume introduced by a point source in forced convective flows is investigated. Advancing to flows in the mixed convective regime in CHAPTER 6, we find a correlation between momentum transport and heat flux. In addition, the spanwise scaling of the identified coherent structures is in agreement with the spatial distribution of the heat flux. In CHAPTER 7 we again address the distribution of a scalar plume introduced by a point source, but now also in the mixed convective regime. The results reveal that transport processes are increased by the presence of buoyancy induced fluid motions. In a numerical approach we apply large eddy simulations (LES) to predict the momentum and scalar field. CHAPTER 8 describes our approach and summarizes the main results. A good agreement between measurements and simulations is obtained, and the numerical results additionally provide more insight into the flow physics. A summary of the results and the main conclusions from them are presented in CHAPTER 9. This thesis is finally concluded with an outlook in CHAPTER 10, which tries to give directions for future research in this area.

Chapter 2

Theoretical Considerations

This chapter provides a description of the fundamental aspects of turbulent flows and of wall influence on turbulence and coherent structures. The governing momentum and scalar equations and their coupling in case of mixed convective flows are introduced in section 2.1. The different scales present in turbulent flows are addressed in section 2.2. Section 2.3 reviews the influence of flat and complex walls on turbulent flows and on coherent structures. The chapter is concluded with the fundamentals of the proper orthogonal decomposition (POD) which is used throughout this thesis to extract coherent structures present in the flow field (section 2.4).

2.1 Governing Equations for Turbulent Flows

The equations describing turbulent flows and heat transfer are given by the conservation of mass, ϱ, momentum, ϱu_i, and thermal energy, $\varrho c_p T$. For the instantaneous motion these equations can be written as

$$\frac{\partial \varrho}{\partial t} + \frac{\partial (\varrho u_j)}{\partial x_j} = 0, \qquad (2.1)$$

$$\frac{\partial (\varrho u_i)}{\partial t} + \frac{\partial (\varrho u_j u_i)}{\partial x_j} = -\frac{\partial p}{\partial x_i} + \frac{\partial}{\partial x_j}\left[\mu\left(\frac{\partial u_i}{\partial x_j} + \frac{\partial u_j}{\partial x_i}\right)\right] + F_i, \qquad (2.2)$$

$$\frac{\partial (\varrho T)}{\partial t} + \frac{\partial (\varrho u_j T)}{\partial x_j} = \frac{\partial}{\partial x_j}\left[\frac{\mu}{\Pr}\left(\frac{\partial T}{\partial x_j}\right)\right], \qquad (2.3)$$

where p is the static pressure, F_i is an external force, μ denotes the dynamic viscosity of the fluid and Pr is the Prandtl number defined as $\Pr = \frac{\mu c_p}{\lambda}$. For flows in the mixed convective regime where buoyancy effects are significant the coupling between the temperature and the velocity field is accomplished by applying the Boussinesq approximation. In this approximation the density changes of the fluid due to a temperature gradient are assumed to be negligible, the resulting buoyancy motion is introduced as external force F_B in the momentum equation, which reads

$$F_B = -\varrho g_i \beta \left(T - T_{ref}\right), \qquad (2.4)$$

where β is the thermal expansion coefficient of the fluid, and g_i the acceleration due to gravity. The working fluid throughout this thesis is water, its properties are tabulated in Table 2.1.

Turbulence is associated with the existence of random fluctuations in the flow field. A way to describe a turbulent flow statistically is to decompose a quantity ϕ into a mean value, $\overline{\phi}$, and a fluctuating part, ϕ'. This procedure is known as Reynolds averaging, where the time–scale

Table 2.1: Properties of water at a temperature of $T = 20$ °C (Wagner and Kruse; 1998).

Property	
Density ϱ (kg/m^3)	1000
Kinematic viscosity ν (m^2/s)	$1 \cdot 10^{-6}$
Thermal conductivity λ (W/(mK))	0.597
Heat capacity c_p (J/(kgK))	4186
Thermal expansion coefficient β (1/K)	$2.07 \cdot 10^{-4}$
Prandtl number Pr	7

of averaging is large compared to the time–scale of turbulence. Applying this procedure to Eqs. 2.1, 2.2, and 2.3 and assuming constant fluid density, ϱ, yields the Reynolds averaged Navier–Stokes equations (RANS)

$$\frac{\partial \overline{u_j}}{\partial x_j} = 0, \qquad (2.5)$$

$$\frac{\partial \varrho \overline{u_i}}{\partial t} + \frac{\partial \varrho \overline{u_j}\, \overline{u_i}}{\partial x_j} = -\frac{\partial \overline{p}}{\partial x_i} + \frac{\partial}{\partial x_j}\left[\mu\left(\frac{\partial \overline{u_i}}{\partial x_j} + \frac{\partial \overline{u_j}}{\partial x_i} - \varrho \overline{u_i' u_j'}\right)\right] + F_i, \qquad (2.6)$$

$$\frac{\partial \varrho \overline{T}}{\partial t} + \frac{\partial \varrho \overline{u_j}\, \overline{T}}{\partial x_j} = \frac{\partial}{\partial x_i}\left[\frac{\mu}{\Pr}\frac{\partial \overline{T}}{\partial x_i} - \varrho \overline{\theta u_i'}\right]. \qquad (2.7)$$

Comparing the averaged to the instantaneous equations the main differences arise from two new unknown terms, the Reynolds stress tensor $\varrho \overline{u_i' u_j'}$ in the momentum balance, and the turbulent heat flux $\varrho \overline{\theta u_i'}$ in the thermal energy equation. To close this system of averaged equations information on the momentum and heat flux are required. This is still an important topic in turbulence modeling and a motivation to assess these quantities experimentally for validation purposes.

2.1.1 Reynolds Stress Equation

Multiplying the equation for the velocity component u_i with u_j, and u_j with u_i, then summing these equations and finally performing Reynolds averaging the transport equation for the Reynolds stress tensor is obtained

$$\frac{\partial \varrho \overline{u_i' u_j'}}{\partial t} + \overline{u_k}\frac{\partial \varrho \overline{u_i' u_j'}}{\partial x_k} = P_{ij} + \Phi_{ij} - \varrho \varepsilon_{ij} - \frac{\partial}{\partial x_k} J_{ijk}, \qquad (2.8)$$

with

$$P_{ij} = -\varrho\left(\overline{u_i' u_k'}\frac{\partial \overline{u_j}}{\partial x_k} + \overline{u_j' u_k'}\frac{\partial \overline{u_i}}{\partial x_k}\right),$$

$$\Phi_{ij} = \overline{p'\left(\frac{\partial u_i'}{\partial x_j} + \frac{\partial u_j'}{\partial x_i}\right)},$$

$$\varrho \varepsilon_{ij} = 2\mu \overline{\frac{\partial u_i'}{\partial x_j}\frac{\partial u_j'}{\partial x_i}},$$

$$J_{ijk} = \overline{p'u'_i}\delta_{jk} + \overline{p'u'_j}\delta_{ik} + \overline{\varrho u'_i u'_j u'_k} - \mu\frac{\partial}{\partial x_k}\overline{u'_i u'_j},$$

where P_{ij} is the production tensor, Φ_{ij} denotes the velocity–pressure gradient, $\varrho\varepsilon_{ij}$ stands for the viscous dissipation, and J_{ijk} denotes the diffusive flux of $\overline{u'_i u'_j}$. In case of $i = j$, one obtains the budget for the turbulent kinetic energy $k = \left(\overline{u'_i u'_i}\right)/2$

$$\frac{\partial k}{\partial t} + u'_j \frac{\partial k}{\partial x_j} = -\overline{\varrho u'_i u'_j}\frac{\partial \overline{u_i}}{\partial x_j} - 2\mu \overline{s'_{ij} s'_{ij}}$$

$$-\frac{\partial}{\partial x_j}\left(\overline{p'u'_j} + \frac{\overline{\varrho u'_i u'_i u'_j}}{2} - 2\mu\overline{u'_i s'_{ij}}\right). \tag{2.9}$$

2.1.2 Turbulent Heat Flux Equation

Multiplying the momentum equation with the fluctuating temperature θ and adding it to the thermal energy equation multiplied with u_i, then after averaging and rearranging the transport equation for the turbulent heat flux is obtained

$$\frac{\partial \overline{\theta u'_i}}{\partial t} + \overline{u_k}\frac{\partial \overline{\theta u'_i}}{\partial x_k} = \mathcal{D}_{\theta i} + \Phi_{\theta i} - P_{\theta i} - \varepsilon_{\theta i}, \tag{2.10}$$

with

$$\mathcal{D}_{\theta i} = \frac{\partial}{\partial x_k}\left(-\overline{\theta u'_i u'_k} + \frac{\overline{\theta p'}}{\varrho}\delta_{ik} + \nu\overline{\theta\frac{\partial u'_i}{\partial x_k}} + a\overline{u'_i\frac{\partial \theta}{\partial x_k}}\right),$$

$$\Phi_{\theta i} = \frac{\overline{p'}}{\varrho}\frac{\partial \theta}{\partial x_i},$$

$$P_{\theta i} = \overline{u'_i u'_k}\frac{\partial \overline{T}}{\partial x_k} + \overline{\theta u'_k}\frac{\partial \overline{u_i}}{\partial x_k},$$

$$\varepsilon_{\theta i} = (\nu + a)\overline{\frac{\partial u'_i}{\partial x_k}\frac{\partial \theta}{\partial x_k}},$$

where $\mathcal{D}_{\theta i}$ denotes the diffusive transport, $\Phi_{\theta i}$ stands for the pressure–temperature correlation, $P_{\theta i}$ is the production term due to combined actions of mean velocity and mean temperature gradient, and $\varepsilon_{\theta i}$ denotes the dissipative correlation.

2.2 Scales of Turbulent Motion

This section introduces in a brief fashion two features of turbulent flows, namely the energy cascade and the smallest scales of turbulent motion.

The idea of the energy cascade is that kinetic energy is introduced into turbulence at the largest scales of motion. This energy is then transferred to successively smaller and smaller scales. At the range of the smallest scales of the turbulent flow the energy is then dissipated by viscous action.

2 Theoretical Considerations

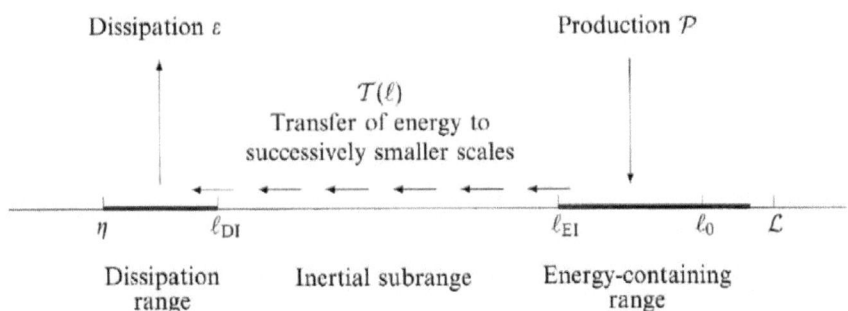

Figure 2.1: Schematic diagram of the scales of turbulent motion and of the energy cascade at very high Reynolds numbers. η denotes the Kolmogorov scale, l_0 the lengthscale of the largest eddies, and $\mathcal{L}$ the characteristic lengthscale of the flow. l_{DI} and l_{EI} mark the boundaries between the dissipation (D) and inertial (I) subranges, respectively between the energy (E) and inertial (I) ranges (Pope; 2000).

Hence, the smallest scales in a turbulent flow are connected to turbulent dissipation, ε. They can be characterized using the Kolmogorov scales (Kolmogorov (1991)), which are the lengthscale η

$$\eta = \left(\frac{\nu^3}{\varepsilon}\right)^{1/4}, \tag{2.11}$$

the time scale τ_η

$$\tau_\eta = \left(\frac{\nu}{\varepsilon}\right)^{1/2}, \tag{2.12}$$

and the velocity scale u_η

$$u_\eta = (\varepsilon\nu)^{1/4}. \tag{2.13}$$

A schematic diagram of the energy cascade and the scales of turbulent motion is depicted in Figure 2.1.

2.3 Wall Influence on Turbulence

In the next sections the wall influence on turbulence and coherent structures are addressed. After a brief summary of the phenomena observed in turbulent boundary layers over flat surfaces, wavy walls are investigated as surfaces with increased complexity.

2.3.1 Flat Wall

In a review by Robinson (1991) a taxonomy of turbulence structures observed in a turbulent boundary layer over a flat surface (low–speed and high–speed refer to perturbations from the mean value at a specific location) is presented:

1. Low–speed streaks in the viscous sublayer.
2. Ejections of low–speed fluid outward from the wall, including lifting low–speed streaks.
3. Sweeps of high–speed fluid inward toward the wall, including inrushes from the outer region.
4. Vortical structures of various forms.
5. Sloping near–wall shear layers, exhibiting local concentrations of spanwise vorticity and $\partial u'/\partial x$.
6. Near–wall pockets, visible in the laboratory as regions swept clean of marked near–wall fluid.
7. Large motions capped by three–dimensional bulges in the outer turbulent/potential interface.
8. Shear–layer backs of large–scale outer–region motions, consisting of sloping discontinuities in the streamwise velocity.

The vortical structures of various forms, mentioned as point 4 of the taxonomy, are often found in literature as arch-, horseshoe, or hairpin–shaped vortical structures.

Studies indicate that the hairpin–shaped structures consist of two counter–rotating quasi–streamwise vortices (e.g. Zhou et al. (1999)). The hairpins do not exclusively posses perfect spanwise symmetry with the two counter–rotating vortex legs of equal strength. In fact, spanwise asymmetric one–sided hairpins have also been observed, and were reported by Zhou et al. (1999). The concept of the turbulent boundary layer as a distribution of hairpin vortices provides explanations for many observed flow features, but is still incomplete.

2.3.2 Wavy Wall

In this thesis the flow between a flat top wall and a complex, wavy bottom surface is considered. The Reynolds number is calculated according to

$$Re_H = \frac{U_B \cdot H}{\nu}, \qquad (2.14)$$

where ν denotes the kinematic viscosity, and H is the height of the channel. The bulk velocity U_B is defined as

$$U_B = \frac{1}{H - y_w} \int_{y_w}^{H} U(x_\xi, y)\, dy \qquad (2.15)$$

where x_ξ denotes an arbitrary x-location and y_w describes the profile of the complex surface. Figure 2.2 shows the coordinate system and schematically illustrates characteristic regions of the flow field in the vicinity of the wavy surface. The coordinate x is directed parallel to the mean flow, y denotes the vertical coordinate direction, and z is the spanwise coordinate direction. The corresponding velocity components are denoted as u, v, and w. Characteristic regions of a flow over waves with separation, reported by Cherukat et al. (1998), and Henn and Sykes (1999), are the separation region (I), and the regions of maximum positive (II) and maximum negative (III) Reynolds shear stress $-\varrho\overline{u'v'}$. Their location with respect to the wavy surface

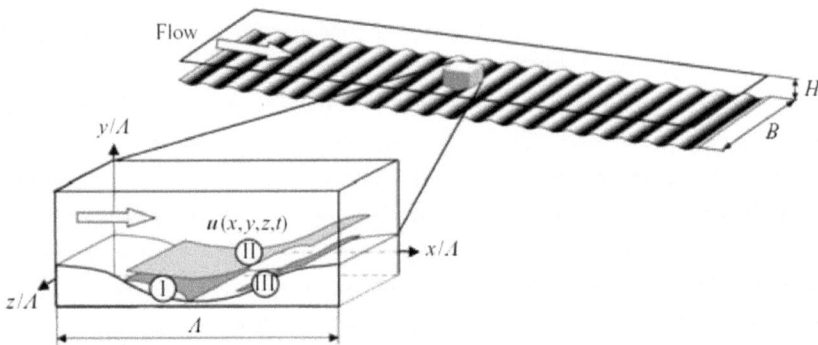

Figure 2.2: Coordinate system and schematic of (I) the separation region, and the regions (II) of maximum positive and (III) maximum negative Reynolds shear stress for a flow situation with separation.

are additionally sketched in Figure 2.2. The geometry of wavy surfaces can be characterized by the dimensionless amplitude $(2a)$–to–wavelength (Λ) ratio

$$\alpha = \frac{2a}{\Lambda}. \qquad (2.16)$$

The flow situations bounded by complex wavy surfaces exhibit characteristic flow features such as the periodicity of the mean flow, partial separation behind the wave crests for large enough α, and a separation zone, sometimes referred to as separation bubble, located in the wave troughs.

A summary of selected experimental studies regarding flow over wavy surfaces in tabulated in Table 2.2. Table 2.3 summarizes selected numerical studies dealing with this flow situation. The most important results and the ambiguities arising from them will be briefly reviewed in the following. The influence of the complexity of the bounding wall, so to speak the wall roughness, on turbulent flows and turbulent flow structures is still an active area of research and investigations. The influence of the wave amplitude on turbulent separated flows has partly been addressed in earlier studies (e.g. Zilker et al. (1977); Zilker and Hanratty (1979)). Raupach et al. (1991) hypothesized that the outer flow has an universal character and is independent of the interaction with the wall. Nakagawa et al. (2003) and Nakagawa and Hanratty (2003) found for the flow over sinusoidal waves that turbulence properties in the outer flow are similar for flat and rough surfaces, when scaled with the friction velocity. Krogstad et al. (1992), Krogstad and Antonia (1994), as well as Krogstad and Antonia (1999) reported an influence of the wall roughness extending well into the outer region of the boundary layer. The results of these studies suggested that the degree of interaction between the wall and the outer region may not be negligible. But as noted in a review by Jiménez (2004), it has been difficult for other investigators to reproduce those findings. Krogstad et al. (1992) also found that the one–point correlation times for all the velocity components are about twice shorter for rough than for smooth boundary layers. Krogstad and Antonia (1994) determined the inclination angle of the two–point correlation function of the streamwise velocity component u between two y locations

2.3 Wall Influence on Turbulence

Table 2.2: Summary of selected experimental studies regarding flow over waves. Setup: CF=channel flow, WT=wind tunnel. $\alpha = 2a/\Lambda$.

Reference	Re_h	α	Setup	Data
Zilker et al. (1977)	7000	0.0125	CF	1D
	6000-32000	0.0312	CF	1D
	7000-32000	0.050	CF	1D
	7000-32000	0.125	CF	1D
	7000-24000	0.200	CF	1D
Thorsness et al. (1978)	5400-30000	0.013	CF	1D
Buckles et al. (1984)	12000	0.200	CF	1D
	12000	0.125	CF	1D
Abrams and Hanratty (1985)	6000-12300	0.014	CF	1D
Frederick and Hanratty (1988)	6400	0.031	CF	1D
	38800	0.050	CF	1D
Kuzan et al. (1989)	48000	0.125	CF	1D
	33000	0.125	CF	1D
	4100	0.200	CF	1D
Hudson et al. (1996)	>3400	0.100	CF	1D
Gong et al. (1996)	3800	0.158	WT	1D
Nakagawa and Hanratty (2001)	46000	0.100	CF	2D
Nakagawa et al. (2003)	46000	0.100	CF	2D
Nakagawa and Hanratty (2003)	3200; 11000	0.100	CF	2D
Günther and Rudolf von Rohr (2003)	500–7300	0.100	CF	2D
Kruse et al. (2003)	4500	0.100	CF	2D
Song and Eaton (2004)	1100-20100	Ramp	WT	2D
Kruse et al. (2006)	11200	0.1; 0.2	CF	2D
Poggi et al. (2007)	150000	0.025	CF	2D
Kuhn et al. (2007)	11200	0.1; 0.2	CF	2D
Wagner et al. (2007)	2300-11200	0.100	CF	2D

Table 2.3: Summary of selected numerical studies regarding flow over waves. Method: DNS=direct numerical simulation, LES=large eddy simulation, CV=contol volume approach. $\alpha = 2a/\Lambda$.

Reference	Re_h	α	Method
Cherukat et al. (1998)	3460	0.100	DNS
Henn and Sykes (1999)	6560-20060	0.031-0.2	LES
Ničeno and Nobile (2001)	175–200	0.25	CV
Stalio and Nobile (2003)	180_τ	Riblets	DNS
Dellil et al. (2004)	6760	0.0-0.1	DNS/$k - \varepsilon$
Tseng and Ferziger (2004)	2400	0.05	LES

as 38° for the rough surface against 10° for the smooth one. As noted in Jiménez (2004), Nakagawa and Hanratty found no change in this quantity. They suggested that this is due to the ambiguity of the advection velocity, arising from the use of particle image velocimetry (PIV), which is a purely spatial procedure. Krogstad and Antonia (1999) state that characterizing the influence of the wall roughness by exclusively evaluating the effect on the mean velocity profiles is inadequate. By comparing measurements over two rough walls with measurements from a smooth wall boundary layer they found that the turbulent characteristics of the flow are significantly affected by the surface geometry.

The research presented in this thesis is driven by the aim to identify the influence of complex surfaces on the flow and on turbulent transport processes, both for isothermal and non–isothermal flow situations. Furthermore we are interested in the coherent structures formed in the vicinity of the complex wall. These coherent, or large–scale structures observed in a shear flow over a wavy wall are identified as spanwise–periodic, streamwise–oriented longitudinal vortices. The influence of the wavy surfaces on this spanwise periodicity and their role in scalar transport processes is also addressed in this thesis. There are two theories in describing this turbulence phenomenon induced by the wavy wall profile, namely the Görtler mechanism and the Langmuir type mechanism which will be reviewed briefly.

Görtler Mechanism The Görtler mechanism explains the generation of coherent structures due to the local curvature of the wall. Boundary–layer instabilities induced by wall curvature are the subject of an extensive review by Saric (1994). The observed flow structures in these configurations are in the form of steady, streamwise–oriented, counter–rotating vortices, commonly called Görtler vortices (Görtler (1940)). The inviscid mechanism was first described by Rayleigh in 1916. According to the *Rayleigh circulation criterion*, in a boundary–layer flow over a concave wall, the radial direction opposes the velocity profile, thus inducing a centrifugal instability. This centrifugal instability then forms the observed coherent structures.

Langmuir Type Mechanism The Langmuir type mechanism describes the observed turbulence structures as being catalysed by the periodicity of the wall profile. In the works of Phillips and Wu (1994), Phillips et al. (1996), and Phillips (2005) these longitudinal flow structures are explained by a Craik-Leibovich type 2 (CL–2) instability. According to this description, the instability is caused by the periodicity of the wavy wall, rather than by its local curvature.

2.4 Proper Orthogonal Decomposition

The proper orthogonal decomposition (POD) is based on correlation functions that are decomposed into a set of orthogonal eigenfunctions. The structural information gathered by this method is obtained from the most dominant eigenfunctions.

2.4.1 Mathematical Background

In the following a set of spatiotemporal data of a scalar quantity $\xi(\mathbf{x},t)$, which may represent a velocity component, $u_i(\mathbf{x},t)$, a temperature, $T(\mathbf{x},t)$, or a concentration of a species, $c(\mathbf{x},t)$, is considered. In general, the two–point correlation function of a scalar quantity $\xi_i(\mathbf{x},t)$ can be written as

$$R_{\xi_i \xi_j}(\mathbf{x},\mathbf{x}',\Delta t) = \overline{\xi'_i(\mathbf{x},t)\xi'_j(\mathbf{x}',t+\Delta t)}, \tag{2.17}$$

where $\xi'_i(\mathbf{x},t)$ denotes the instantaneous fluctuations of the ith component of the scalar quantity at positions $\mathbf{x}$, $\mathbf{x}'$ and times t, $t + \Delta t$.

Following the approach of Moin and Moser (1989) and restricting to spatial variations, the two–point correlation function for a turbulent flow with one direction of flow inhomogeneity and two homogeneous directions is written as

$$R_{\xi_i \xi_j}(\Delta x, y, y', \Delta z) = \left\langle \xi'_i(x,y,z,t) \cdot \xi'_j(x+\Delta x, y', z+\Delta z, t) \right\rangle_{xt}, \tag{2.18}$$

where ξ'_i ($i = 1,2,3$) denotes the instantaneous fluctuations of the scalar quantity, and $\langle \cdot \rangle_{xt}$ denotes ensemble averaging over the homogeneous direction, here the streamwise coordinate x, and over the time coordinate t as well.

Since the flow field data are two–dimensional the correlation function can be written for every plane of measurement. In case of the (x,y)–plane, where x is the homogeneous direction and y is the inhomogeneous direction, it reads

$$R_{\xi_i \xi_j}(\Delta x, y, y') = \left\langle \xi'_i(x,y,t) \cdot \xi'_j(x+\Delta x, y', t) \right\rangle. \tag{2.19}$$

The aim of the proper orthogonal decomposition is to describe a given flow field through a minimal number of deterministic modes. According to Berkooz et al. (1993) this leads to the following equation

$$\int_I R_{\xi\xi}(\mathbf{x},\mathbf{x}')\Pi_i(\mathbf{x}')d\mathbf{x}' = \lambda_i \Pi_i(\mathbf{x}), \tag{2.20}$$

where $R_{\xi\xi}$ denotes the cross–correlation function between the points $\mathbf{x}$ and $\mathbf{x}'$, and I denotes the interval on which the functions are defined. The solution of Eq. 2.20 forms a complete set of square–integrable, which have in physical terms a finite kinetic energy with associated eigenvalues λ_i and eigenfunctions Π_i. With this, any ensemble may be reproduced by a modal decomposition in the eigenfunctions, which means by a series of orthogonal functions with random, un–correlated coefficients $a_i(t)$:

$$\xi(\mathbf{x},t) = \sum_{i=1}^{\infty} a_i(t) \Pi_i(\mathbf{x}). \tag{2.21}$$

The cross–correlation tensor can be written in terms of Π_i itself:

$$R_{\xi\xi}(\mathbf{x},\mathbf{x}') = \sum_{i=1}^{\infty} \lambda_i \Pi_i(\mathbf{x}) \Pi_i^T(\mathbf{x}'). \tag{2.22}$$

Finally for the energy follows:

$$E = \langle (\xi,\xi) \rangle = \int_I R(\mathbf{x},\mathbf{x}) d\mathbf{x} = \sum_{i=1}^{\infty} \lambda_i. \tag{2.23}$$

This means that the eigenvalues provide the energy contribution of the various eigenfunctions. The mathematical formalism of POD is fully described by Eq. 2.20 and Eq. 2.21. An interesting special case occurs when the correlation tensor only depends on the difference between two coordinates. In this case the POD modes correspond to Fourier modes.

2.4.2 Method of Snapshots

The method of snapshots is based on the POD and was introduced by Sirovich (1987). In the following, discrete times t_i with $i = 1, \ldots, M$, and $1, \ldots, N$ discrete locations within a two–dimensional plane are considered. Thus, a spatiotemporal set of data can be written as the following $N \times M$ matrix:

$$\mathbf{X} = \{\mathbf{X}_i\}_{i=1}^{M} = \begin{bmatrix} \xi_{11}, \xi_{12}, \ldots, \xi_{1M} \\ \xi_{21}, \xi_{22}, \ldots, \xi_{2M} \\ \vdots \\ \xi_{N1}, \xi_{N2}, \ldots, \xi_{NM} \end{bmatrix} \tag{2.24}$$

with $\mathbf{X}_i = [\xi_1, \xi_2, \ldots, \xi_N]^T$.
The mean is computed by

$$\overline{\mathbf{X}} = \frac{1}{M} \sum_{i=1}^{M} \mathbf{X}_i. \tag{2.25}$$

For the fluctuations it then follows that

$$\mathbf{X}'_i = \mathbf{X}_i - \overline{\mathbf{X}}, \quad i = 1, \ldots, M. \tag{2.26}$$

Applying the method of snapshots, a $M \times M$ covariance matrix $\mathbf{C}$ can be computed, which reads

$$\mathbf{C}_{ij} = \langle \mathbf{X}'_i \mathbf{X}'_j \rangle, \quad i,j = 1, \ldots, M, \tag{2.27}$$

2.4 Proper Orthogonal Decomposition

where $\langle \cdot, \cdot \rangle$ is the Euclidean inner product. Since the covariance matrix is symmetric, its eigenvalues, λ_i, are non–negative, and its eigenvectors, ϕ_i, $i = 1, \ldots, M$, form a complete orthogonal set. The orthogonal eigenfunctions are thus defined as:

$$\Pi^{[k]} = \sum_{i=1}^{M} \phi_i^{[k]} \mathbf{X}'_i, \qquad k = 1, \ldots, M, \tag{2.28}$$

where $\phi_i^{[k]}$ is the ith component of the kth eigenvector. As already given by Eq. 2.23, the total energy E is obtained through summation of the eigenvalues:

$$E = \sum_{i=1}^{M} \lambda_i.$$

The fractional contribution of each eigenfunction to the total energy is given by the fractional contribution of its associated eigenvalue:

$$\frac{E_k}{E} = \frac{\lambda_k}{E}. \tag{2.29}$$

And as already mentioned above, any sample field $\mathbf{X}_j$ can be reconstructed by using the eigenfunctions $\Pi^{[i]}$:

$$\mathbf{X}_j = \overline{\mathbf{X}} + \sum_{i=1}^{M} a_i \Pi^{[i]}, \tag{2.30}$$

where the coefficients a_i are computed from the projection of the sample vector $\mathbf{X}'_j$ onto eigenfunction $\Pi^{[i]}$:

$$a_i = \frac{\mathbf{X}'_j \cdot \Pi^{[i]}}{\Pi^{[i]} \cdot \Pi^{[i]}}. \tag{2.31}$$

By using only the first K, where $K < M$, most energetic eigenfunctions it is possible to construct an approximation to the data

$$\mathbf{X}_j = \overline{\mathbf{X}} + \sum_{i=1}^{K} a_i \Pi^{[i]}. \tag{2.32}$$

Chapter 3

Experiments and Experimental Techniques

In this chapter the experimental facility and the applied experimental techniques are described. The experiments are conducted in a closed–loop channel facility, where the bottom wall in the test section can be interchanged between different complex surfaces composed of sinusoidal waves (section 3.1). These surfaces can be heated with an integrated electrical heating foil to address heat transfer processes. To measure the velocity field particle image velocimetry (PIV) is used (section 3.2). The scalar fields (concentration of a species or the fluid temperature) are assessed by means of laser induced fluorescence (LIF) (section 3.3). Both techniques are combined to enable simultaneous measurements to address the structure of turbulent scalar fluxes (section 3.4).

3.1 Channel Facility

The measurements are carried out in the channel facility depicted in Figure 3.1. The working fluid is de–ionized and filtered water, the channel facility is made of both regular and anodized aluminium, PVC, and Schott BK–7 glass. All parts are positioned in a welded stainless steel frame.

A frequency controlled stainless steel pump (Egger T 62–80 HF6, (10) in Figure 3.1) draws the water from an intake, installed at the bottom of the reservoir tank (9), and pumps it through a PVC tube with an inner diameter of 50 mm (11). Following the PVC tube, a 2.25 m long diffusor with a maximum opening angle of $\alpha = 1.9°$ is installed (12). The cross–section of the diffusor changes gradually from circular to a rectangular one of 360 mm × 200 mm. The flow direction is changed by the turning vanes (1). Following the turning vanes the flow passes a honeycomb (2), made of carbon–fibre reinforced plastics having a hexagonal cell structure. The length–to–cell–size ratio of the honeycomb is seven. At the entrance to the rectangular channel (3), the cross–sectional area is reduced by factor 6.7. In addition a boundary layer trip, protruding 1 mm into the flow from each of the four walls, is installed. This ensures a fully developed turbulent channel flow by disturbing the boundary layer of the flow, which now enters the test section (4)–(8) consisting of a flat top wall and complex bottom surface. The full height of the test section, H, is 30 mm, its width, B, is $12H$ (aspect ratio $B/H = 12:1$). The flat–wall entrance section (3) is 67 channel heights long with a wall thickness of 6 mm. The test section (5), consisting of a flat top wall and a complex bottom wall, has a length of 72 channel heights. The recirculation system and the flat–walled entrance section are depicted in Figure 3.2.

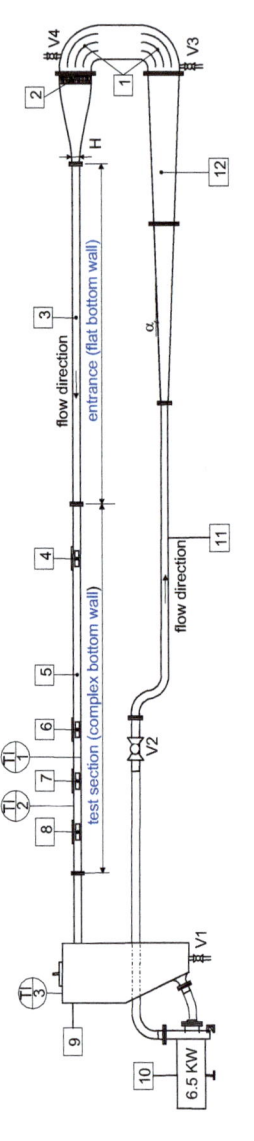

Figure 3.1: Facility with the channel sections and the recirculation system: (1) turning vanes, (2) honeycomb, acceleration section, (3) flat-walled entrance section, (4)–(8) optical view ports, test section, (9) reservoir, (10) frequency controlled stainless steel pump, (11) PVC tube, (12) diffusor ($\alpha = 1.9°$).

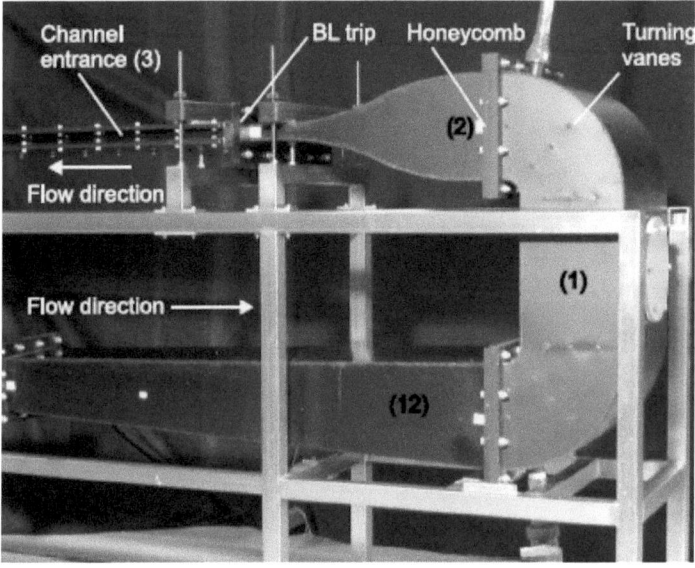

Figure 3.2: Flat–walled entrance section of the channel and recirculation system: (1) turning vanes, (2) honeycomb, acceleration section, (3) flat–walled entrance section, (12) diffusor ($\alpha = 1.9°$).

3.1.1 Test Section

Optical access for particle image velocimetry and laser induced fluorescence measurements is provided at four streamwise positions through viewing ports, depicted in Figure 3.3. The viewing ports are made of optical grade Schott BK–7 glass, and are positioned on both sidewalls (thickness 5 mm), and on the flat top wall (thickness 7 mm). Measurements are performed for the developed flow after a distance corresponding to 50 channel heights. The side windows provide a maximum area of view (AOV) of $3H$ (streamwise)×$1.2H$ (wall–normal), the maximum AOV for the top windows is $3.3H$ (streamwise)×$3.3H$ (spanwise).

The investigated bottom surfaces are three different sinusoidal walls with amplitude-to-wavelength ratios of $\alpha = 2a/\Lambda = 0.1$ ($\Lambda = 30$ mm), $\alpha = 0.2$ ($\Lambda = 30$ mm), and $\alpha = 0.2$ ($\Lambda = 15$ mm), and a surface consisting of two superimposed sinusoidal waves with $\alpha = 0.1$ ($\Lambda = 30$ mm). The two types of complex surfaces and the coordinate system used are depicted in Figure 3.4. The streamwise coordinate direction is denoted by x, the vertical direction by y, and z is the spanwise direction. The corresponding velocity components are denoted as u, v, and w. The profile of the wavy walls is described by

$$y_w(x) = a\cos\left(\frac{2\pi x}{\Lambda}\right), \tag{3.1}$$

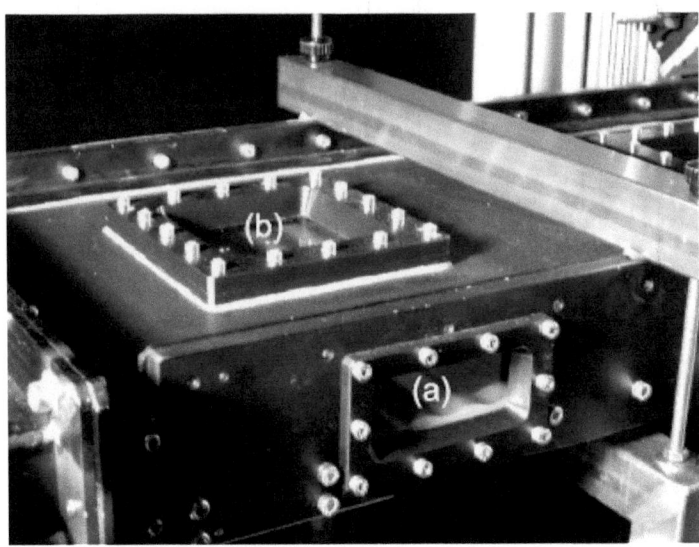

Figure 3.3: Optical access through (a) a pair of windows on the sidewalls and (b) one window on the flat top wall.

the superposition of the two sinusoidal waves is given by

$$y_w(x,z) = a \cos\left(\frac{2\pi x}{\Lambda}\right) \cos\left(\frac{2\pi z}{\Lambda}\right). \qquad (3.2)$$

Figure 3.5 depicts the parameters to describe the complex surfaces. The flow over the wavy walls is homogeneous in the spanwise direction, whereas for the flow over the superimposed waves inhomogeneous flow conditions are achieved.

For the non–isothermal flow situations the complex bottom surface in the test section is made of aluminum and is heated with an embedded electrical heating foil ($\dot{q} = 1800 \text{ W/m}^2$). Two temperature sensors are integrated in the top and bottom wall to control and to record the temperature during the measurements at the measurement location. A heat exchanger is incorporated in the reservoir (9) to remove the heat from the fluid after passing the test section and to thereby ensure a constant entry temperature into the test section. These non–isothermal measurements have only been performed for the wavy wall with $\alpha = 0.1$ ($\Lambda = 30$ mm) and for the superimposed waves with $\alpha = 0.1$ ($\Lambda = 30$ mm).

3.1.2 Experiments

The measurements are conducted in three perpendicular planes, namely the (x,y)–, (x,z)–, and the (y,z)–plane (the coordinate system used is depicted in Figure 3.4). The range of Reynolds numbers considered is between $\text{Re}_H = 1000$ and $\text{Re}_H = 40000$. The Reynolds number can be continuously adjusted by the frequency control of the pump, where the maximum obtainable

3.1 Channel Facility

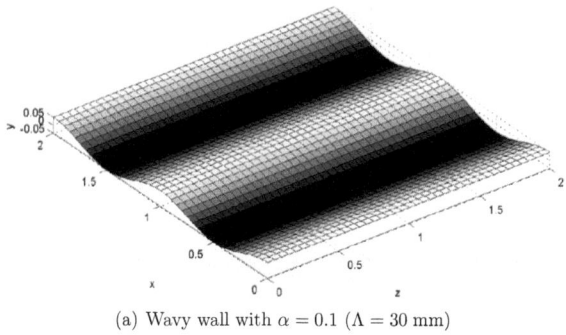

(a) Wavy wall with $\alpha = 0.1$ ($\Lambda = 30$ mm)

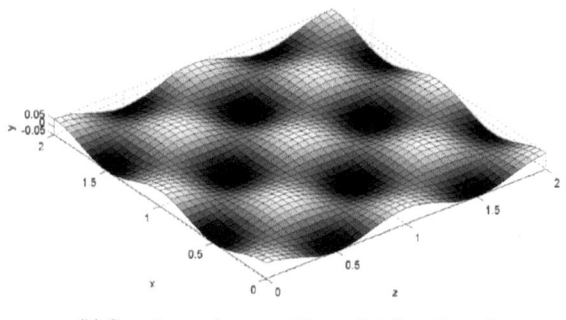

(b) Superimposed waves with $\alpha = 0.1$ ($\Lambda = 30$ mm)

Figure 3.4: Profiles of the complex surfaces used as bottom wall in the test section.

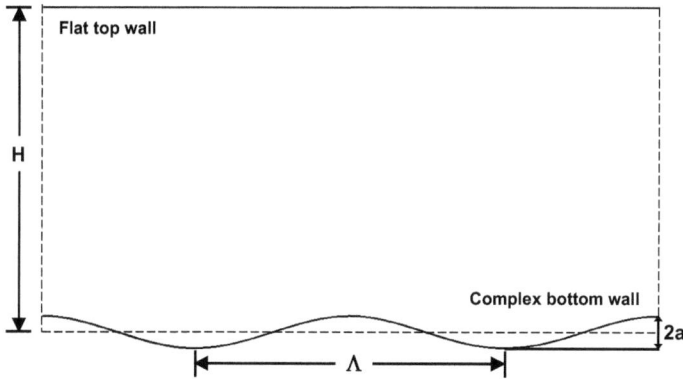

Figure 3.5: Sketch of the test section with the parameters to describe the complex surfaces.

Figure 3.6: Water prism to correct for optical distortions when measuring in the (y,z)–plane.

frequency is 50 Hz, which corresponds to a Reynolds number of 40000, calculated with the total height of the channel, H, and the bulk velocity, U_B.

The light sheet is adjusted to the middle of the flow channel ($z = 0.5B$) for the measurements in the (x,y)–plane. To measure in the (x,z)–plane the light sheet is aligned in a plane above the complex surface parallel to the flat top wall. The vertical coordinate y/H of the light sheet is variable and can be positioned by a traversing system which is connected to the laser, the laser optics, and the imaging optics. The accuracy of this vertical adjustment is approximately 10 μm. For the measurements in the (y,z)–plane the light sheet is aligned perpendicular to the top wall and intersects with the wavy wall in the wave trough. To allow distortion–free imaging through the optical viewing port at the top wall of the channel, a second optical grade glass window is positioned parallel to the imaging optics and the volume between the two glass windows is filled with water. This configuration is often referred to as water prism (Figure 3.6). The deformation of the recorded images due to the fact that the light sheet and the imaging optics are not parallel is accounted for by applying a perspective mapping function.

To study the structure of a dye plume injected at the complex surface a flush mounted low momentum point source is integrated in the bottom wall located at the wave crest of the wall profile. This point source consists of a canula with an inner diameter of 0.8 mm which is connected to a calibrated syringe pump, the inlet flow rate is adjusted to 0.25 ml/min, a value which is found not to perturb the mean flow (Wagner et al. (2007)).

3.2 Particle Image Velocimetry

Particle image velocimetry (PIV) is performed to examine the spatial variation of the velocity components in the plane of measurement. PIV is a nonintrusive measuring technique which is based on multiple recordings of particle images tracing the flow field in the region of interest.

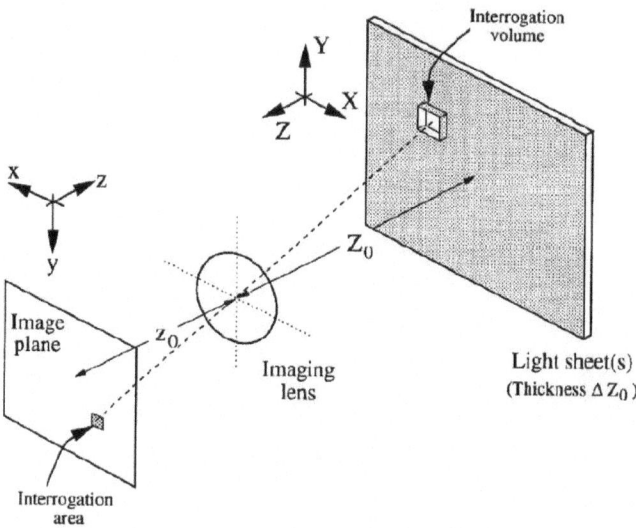

Figure 3.7: Schematic representation of the projection of the interrogation volume onto the image plane (Raffel et al.; 1998).

These multiple realizations are then statistically analyzed to yield velocity fields. For a comprehensive literature survey we refer to Adrian (1991); Westerweel (1993, 1997); Raffel et al. (1998).

The flow is seeded with small tracer particles, which are described more detailed in the next section 3.2.2. These particles are assumed to follow the fluid flow without interference, which means that the particle velocity, which is determined by PIV, is identical to the local fluid velocity. The particles pass a light sheet generated by a laser and the light scattered from their surfaces is recorded on the CCD chip of a digital camera. The light sheet is illuminated twice by two laser pulses, giving a pair of two separate images. The time delay between the two pulses depends mainly on the mean flow velocity, as the particle shift between the two images should be in the range of 5 to 10 pixels. The local displacement vectors of the particles can be obtained by a local cross–correlation of two subsequent images. For this cross–correlation the digital PIV recording is subdivided into smaller subareas, the so called interrogation areas. The local displacement vectors are determined in each interrogation area of the image. Knowing the time delay between the laser pulses, the velocity field in the two–dimensional plane is given by the particle displacement vectors. More details about the evaluation procedure and the accuracy of PIV are given in section 3.2.3.

3.2.1 Geometric Imaging

Figure 3.7 depicts schematically the projection of the scattered light by the tracer particles onto the image plane. The optical system can be represented by a spherical lens with focal length f and aperture diameter D_a, the image plane refers then to the CCD chip. Z_0 denotes the distance between the object and the lens, z_0 the distance between the lens and the image plane. These distances, Z_0 and z_0, satisfy the geometrical law

$$\frac{1}{Z_0} + \frac{1}{z_0} = \frac{1}{f}. \tag{3.3}$$

The image magnification is defined as

$$M = \frac{z_0}{Z_0}. \tag{3.4}$$

It is important to note, that the image of a distant point source, even if an abberation–free lens is used, does not appear as a distinct point on the image plane, but forms a Fraunhofer diffraction pattern. According to Adrian (1991), the particle image diameter for diffraction limited imaging can be calculated from

$$d_e = \sqrt{(Md_P)^2 + (2.44 f_\# (M+1) \lambda)^2}, \tag{3.5}$$

where λ denotes the wavelength of the laser emitted light, and $f_\#$ is the f–number of the camera objective, defined by

$$f_\# = \frac{f}{D_a}.$$

The focal depth of the camera objective, δ_z, is given by

$$\delta_z = 4.88\lambda \left[f_\# \left(1 + \frac{1}{M}\right) \right]^2. \tag{3.6}$$

The latter considerations are important for the setup of the optical system used in experiments. The particle image diameter recorded on the image plane (CCD chip) should at least cover 2–3 pixels to be correctly recognized by the cross–correlation algorithm.

3.2.2 Tracer Particles

Properties of Tracer Particles

The following properties of the tracer particles, being immersed in the fluid, are of great importance for PIV measurements:

- particle density ϱ_P
  The particle density, ϱ_P, has great influence on the way the particles interact with the fluid flow, namely if the particles move with the local fluid velocity or if the particles are equally buoyant. Desirable would be a density equal to the one of the working fluid.

- particle diameter d_P and particle diameter distribution
  The mean diameter of the particles should be small enough not to affect the smallest scales in the flow. On the other hand, the smaller the particles are, the smaller is their light scattering efficiency. A narrow particle diameter distribution would also be desirable. The particle diameter and the particle diameter distribution also have an effect on the way the particles interact with the fluid flow, i.e. that the particles move with the local flow velocity.

- light scattering properties
  The light scattering properties of a solid particle depend on the particle diameter, but also on the shape of the particle. Ideally, the particles would be of spherical shape.

- seeding density
  Since the cross–correlation of the particle images to determine the particle displacement is a statistical method the seeding density is an important factor. The number of particle images per interrogation area should lie in the range between 5 and 10.

For the measurements described in this thesis hollow glass spheres ($\varrho_P = 1.03$ g/cm^3; $d_P = 10$ μm) and polyamide particles ($\varrho_P = 1.01$ g/cm^3; $d_P = 50$ μm) were used as seeding particles. These particles possess all the important properties described above.

Relative Motion of Particles in a Turbulent Flow

The accuracy of PIV measurements relies on tracer particles moving with the local fluid velocity. Hjelmfelt and Mockros (1966) investigated the relative motion between particles and a fluid theoretically. In their derivation they made the following assumptions:

- homogeneous and stationary turbulence

- spherical particles with a mean diameter, d_P, smaller than the smallest eddies

- small Reynolds numbers based on the relative velocity of the fluid and the particles (i.e. Stokes flow)

- sufficiently dilute suspension, such that particle–particle interactions can be neglected

With these assumptions the equation of motion is given by the Basset equation:

$$\frac{\pi d_P^3}{6}\left(\varrho_P + \frac{1}{2}\varrho\right)\frac{du_{rel}}{dt} = -3\pi\nu\varrho u_{rel} - \frac{3}{2}d_P^2\varrho\sqrt{\pi\nu}\int_{t_0}^{t}\frac{du_{rel}/d\xi}{\sqrt{t-\xi}}d\xi, \tag{3.7}$$

where u_{rel} is the relative velocity of the fluid and the particle, $u_{rel} = u_P - u$.

The acceleration of the fluid adds a pressure force on the spherical particles, resulting in a term $\frac{1}{6}\pi d_P^3\varrho\,(du/dt)$. Furthermore, the following assumptions can be made:

$$\frac{d_P^2}{\nu}\frac{\partial u}{\partial x} \ll 1$$

$$\frac{u}{d_P^2}\frac{1}{\partial^2 u/\partial x^2} \gg 1$$

Taking all this into account, Eq. 3.7 becomes

$$\frac{du_P}{dt} + a\cdot u_P + c\cdot\int_{t_0}^{t}\frac{du_P/d\xi}{\sqrt{t-\xi}}d\xi = a\cdot u + b\cdot\frac{du}{dt} + c\cdot\int_{t_0}^{t}\frac{du/d\xi}{\sqrt{t-\xi}}d\xi, \tag{3.8}$$

in which

$$a = \frac{18\nu}{\{(\varrho_P/\varrho)+\frac{1}{2}\}d_P^2},$$

$$b = \frac{3}{2\left\{(\varrho_P/\varrho) + \frac{1}{2}\right\}},$$

$$c = \frac{9}{\left\{(\varrho_P/\varrho) + \frac{1}{2}\right\}}\sqrt{\frac{\nu}{\pi}}.$$

To describe the relative motion between the particle and the fluid, Hjelmfelt and Mockros (1966) introduced two parameters, the amplitude ratio, η, and the phase angle, β:

$$\eta = \sqrt{(1+f_1)^2 + f_2^2}, \tag{3.9}$$

$$\beta = \tan^{-1}\left\{\frac{f_2}{1+f_1}\right\}. \tag{3.10}$$

The analytical expressions for f_1 and f_2 can be written in terms of the circular frequency of the fluid/particle motion, ω:

$$f_1 = \frac{\omega\left(\omega + c\sqrt{\pi\omega/2}\right)(b-1)}{\left(a + c\sqrt{\pi\omega/2}\right)^2 + \left(\omega + c\sqrt{\pi\omega/2}\right)^2},$$

$$f_2 = \frac{\omega\left(a + c\sqrt{\pi\omega/2}\right)(b-1)}{\left(a + c\sqrt{\pi\omega/2}\right)^2 + \left(\omega + c\sqrt{\pi\omega/2}\right)^2}.$$

The Stokes number, N_S, is defined by

$$N_S = \sqrt{\frac{\nu}{\omega d_P^2}}. \tag{3.11}$$

Figure 3.8 shows the solutions of the Basset equation for tracer particles with a mean diameter of $d_P = 50$ μm at various density ratios (ϱ_P/ϱ=1.05, 1.1, 1.2, 1.5, 2, and 3). The particle diameter chosen for the calculation represents the size of the largest tracer particles used in the experiments, which are likely to deviate most from the local fluid motion. For a turbulent channel flow at the considered Reynolds numbers, the frequency of velocity fluctuations lies in the range of approximately 50 – 100 Hz (e.g. Günther et al. (1998)). This results in a Stokes number of $N_S = 2.8$ (50 Hz), respectively $N_S = 2.0$ (100 Hz). At these conditions Figure 3.8 shows, that there is nearly no deviation of the particle motion from the fluid motion. This is even true for considerably large density ratios.

3.2.3 Measurement Accuracy

The measurement accuracy of PIV is influenced by the particle image diameter, the particle image density, the particle image displacement and the velocity gradient within the interrogation area. Thus the total error can be subdivided into errors resulting from the image processing (interrogation scheme) and errors resulting from the image recording. Taking into account the findings in the previous sections the errors due to image recording are negligible and can be estimated to be 0.01% (Kruse (2005)).

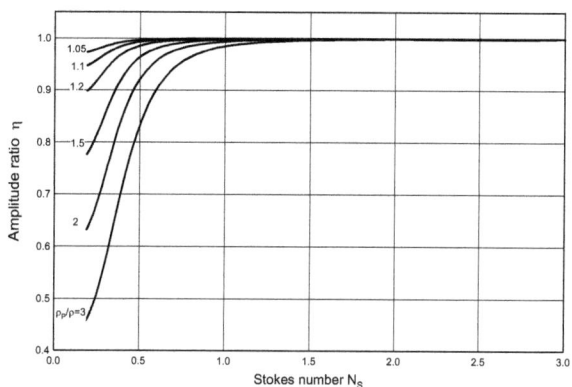

(a) Amplitude ratio η versus Stokes number N_S.

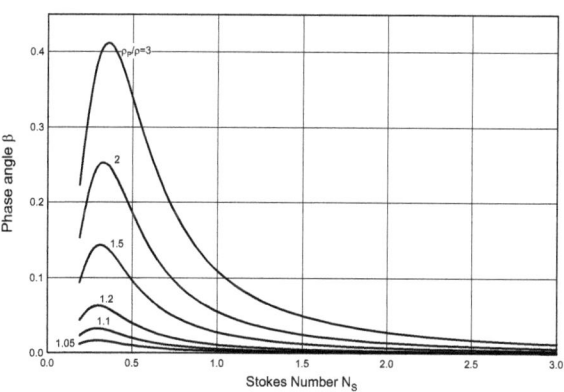

(b) Phase angle β versus Stokes number N_S.

Figure 3.8: Solutions of the Basset equation for a mean particle diameter of $d_P = 50$ μm and the density ratios ϱ_P/ϱ=1.05, 1.1, 1.2, 1.5, 2, and 3.

The image processing is subdivided into the following steps (Scarano and Riethmuller (2000); Scarano (2002)): First the PIV images are interrogated with a basic cross-correlation method on a coarse grid (interrogation area 32 × 32 pixels^2). The results are afterwards validated, spurious vectors are locally filtered and interpolated in order not to affect the further processing steps. The results of this first step are then used to build a predictor displacement field over all image pixels. According to this predictor spatial distribution the two images are deformed, for the sub-pixel shifting B–splines are used for interpolation. These deformed images are then interrogated (interrogation area 32 × 32 pixels^2) with an adaptive cross–correlation, the resulting velocity field is again locally filtered to remove spurious vectors. This procedure of image deformation is then repeated with a predefined refinement of the interrogation areas (interrogation area 16 × 16 pixels^2). The subsequent interrogation yields a displacement field with a much finer resolution which is then further analyzed following local filtering and interpolation.

Due to the considered interrogation algorithm the error level for the image processing is estimated to be below 10^{-2} pixels. Thus the uncertainty of the PIV measurements resulting from the experimental setup, image acquisition and image processing is estimated to be in the order of 1%.

3.2.4 Spatial Resolution

Another important aspect when discussing results obtained by PIV is to determine the spatial resolution of the measurements, which affects the scales of the turbulent flow resolved by PIV.

As already mentioned in section 2.2, the smallest scales of a turbulent flow can be estimated by calculating the Kolmogorov length scale η, and the Kolmogorov time scale τ_η respectively:

$$\eta = \left(\frac{\nu^3}{\varepsilon}\right)^{1/4},$$

$$\tau_\eta = \left(\frac{\nu}{\varepsilon}\right)^{1/2}.$$

The viscous dissipation, ε, can be calculated by

$$\varepsilon = \frac{f\overline{u}^3}{2D}, \tag{3.12}$$

where D is the hydraulic diameter, and f is a friction coefficient. For the turbulent case, f can be estimated by

$$f = 0.316 Re^{-0,25}. \tag{3.13}$$

Figure 3.9 depicts the Kolmogorov scales subject to the Reynolds number.

The smallest resolvable length scale by PIV, l_r, can be estimated when the size of the interrogation area, the size of the CCD chip and the size of the area of view (AOV) mapped onto the CCD chip are known. The average spatial resolution of PIV measurements reported in this thesis lies in a range of 400 μm. Comparing this to results obtained in Figure 3.9 it follows that the Kolmogorov scales are about one order of magnitude smaller than the resolvable length scales. However, the range of resolvable scales by PIV is sufficient to address turbulence phenomena like e.g. organized motions, turbulence production processes, and turbulent transport processes.

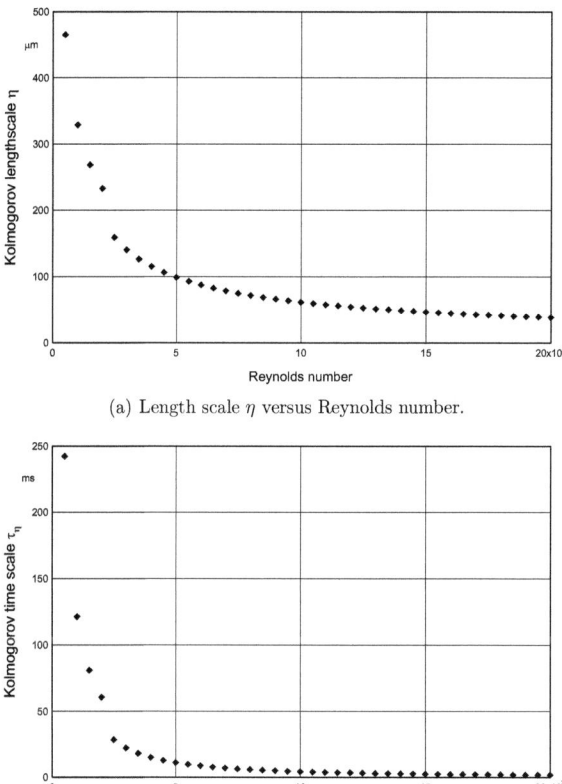

(a) Length scale η versus Reynolds number.

(b) Time scale τ_η versus Reynolds number.

Figure 3.9: Spatial and temporal Kolmogorov scales subject to the Reynolds number.

3.3 Laser Induced Fluorescence

Scalar measurements are performed with the method of laser induced fluorescence (LIF). The planar LIF technique allows two–dimensional measurements of instantaneous scalar fields, i.e. the concentration c of a tracer species or of the temperature field T. In LIF, a laser source is used for the excitation of a fluorescent dye present in the flow field. A part of the laser light is absorbed and a part of this absorbed energy from the laser source is spontaneously re–emitted. The ratio of the total energy emitted per quantum of energy absorbed by the molecule is called the quantum efficiency, ϕ. The fluorescence energy I emitted per unit volume is defined as

$$I = I_0 \phi \varepsilon c, \tag{3.14}$$

where I_0 is the incident light intensity, c is the concentration of the fluorescent dye in the measurement location, and ε is an absorption coefficient.

A flashlamp pulsed dual Nd:YAG laser provides the pulsed light source for the planar laser–induced fluorescence technique. Karasso and Mungal (1996, 1997) showed that the Nd:YAG laser is a suitable laser for the PLIF technique. For instantaneous realizations of scalar fields, the image generation time for PLIF has to be smaller than any relevant fluid mechanical time-scale. The fluorescence lifetime can be estimated to be in the order of 15 ns and fluorescence emission of typical tracer dyes is less than 5 ns (Karasso and Mungal (1997)). The time–scales for turbulent scalar mixing processes can be estimated as of $\mathcal{O}(10^{-6}$ s), thus the use of pulsed Nd:YAG lasers for the PLIF technique is justified.

For laboratory investigations there are two possibilities to apply LIF, namely by using one fluorescent dye (one–color LIF) or by having two fluorescent dyes diluted in the measurement volume (two–color LIF). Both methods are described in more detail in the following sections.

3.3.1 One–color Laser Induced Fluorescence

Throughout this thesis one–color LIF is employed to measure the concentration field of a scalar emanating from a point source. In spite of the advantages of using pulsed Nd:YAG lasers as light source for LIF there are two major drawbacks:

- Inhomogeneities in the laser light sheet due to non-uniformities in the beam profile and due to the imperfection of the optics to generate a light sheet

- A pulse–to–pulse variation of the emitted beam energy in the order of $\mathcal{O}(1\%)$

These disadvantages have to be accounted for in the calibration procedure. The inhomogeneities in the laser light sheet are compensated by applying a single pixel calibration, i.e. a calibration function is generated for each pixel of the CCD chip of the LIF camera. To account for the pulse–to–pulse variation of the laser beam there are two possible methods. The first is to measure the pulse energy online during the measurement and record it for further correction of the data. The second method is to calibrate large ensemble averaged images instead of single images. Throughout this thesis the latter method is applied.

As can be derived from Eq. 3.14, by accounting for the changes in the incident light intensity and calibrating for ϕ and ε it is possible to relate the emitted light intensity from the tracer dye to the local tracer concentration. It has to be noted that we do not calibrate for the quantum efficiency ϕ and the absorption coefficient ε separately, but for the product $\phi \varepsilon$.

3.3.2 Two–color Laser Induced Fluorescence

We use two–color LIF to asses the temperature field of the fluid as proposed by Sakakibara and Adrian (1999, 2004). For some organic dyes, the quantum efficiency ϕ is temperature dependent. Thus by keeping constant the tracer concentration c and accounting for the changes in the incident light intensity I_0 and calibrating the optics (ε) the emitted light intensity provides information about the local fluid temperature. In contrast to one–color LIF the inhomogeneities of the laser light sheet and the pulse–to–pulse variations are compensated by using a second fluorescent dye which is homogeneously diluted in the measurement volume but is not sensitive to temperature changes. Thus by computing the ratio γ of the emitted light intensities of the temperature sensitive and the temperature insensitive dye the corrected temperature information can be derived. In addition we also apply the aforementioned methods of single pixel calibration and ensemble averaging.

3.3.3 Properties of Tracer Dyes

In this thesis Rhodamine B is used as fluorescent dye for the measurement of the concentration field with one–color LIF, and a combination of Rhodamine B and Rhodamine 110 for the temperature measurements with two–color LIF. In this case Rhodamine B will be the temperature sensitive dye, whereas Rhodamine 110 shows no changes in emitted light intensity with respect to temperature.

For the concentration measurements of a tracer dye injected by a point source it is important to estimate the rate of molecular diffusion compared to diffusion through turbulent mixing. This is achieved by calculating the Schmidt number Sc, which represents the ratio of momentum diffusivity (i.e. fluid viscosity ν) and diffusivity of a species D. The molecular diffusion coefficient (calculated with the Wilke–Chang equation) of Rhodamine B in water at a temperature of 20 °C is $D = 3.6 \cdot 10^{-10}$ m^2/s, which yields a Schmidt number of Sc $\approx$ 2800. This means that the studied scalar transport phenomena are dominated by turbulent momentum transport rather than by molecular diffusion of the species. The instantaneous concentration fields provide the information about the mixing due to the influence of turbulent motions in the fluid.

Another important aspect is the use of both fluorescent dyes in combination with a Nd:YAG laser as excitation light source. The absorption spectra of Rhodamine B, respectively Rhodamine 110, diluted in water are depicted in Figs. 3.10 and 3.11. These figures show that Rhodamine B exhibits good absorption properties at an excitation wavelength of 532 nm, whereas for Rhodamine 110 the excitation wavelength is situated at the tail of the absorption peak. Thus less quantum efficiency is expected for Rhodamine 110. To test the suitability of this dye combination we measured the emission spectra of a mixture of Rhodamine B and Rhodamine 110 (ratio 1:1) excited at a wavelength of 532 nm which is depicted in Figure 3.12. It is observed that the emission peaks of both dyes are pronounced and clearly distinguishable. The wavelength of maximum emission of Rhodamine B is 575 nm, for Rhodamine 110 a value of 520 nm is found. Thus we conclude that this dye combination is suitable for temperature measurements with Nd:YAG lasers. The phenomena that Rhodamine 110 emits light at a shorter wavelength compared to the excitation wavelength is known in literature as Anti–Stokes shift, e.g. Bojarski et al. (1977). The properties of Rhodamine B and Rhodamine 110 are summarized in Table 3.1.

In addition the temperature dependence of Rhodamine B needs to be investigated. To address

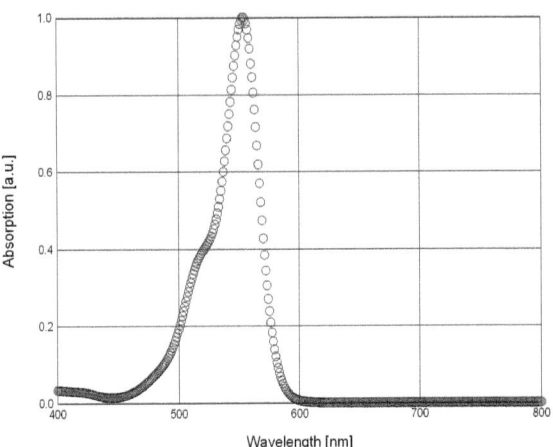

Figure 3.10: Absorption spectrum of Rhodamine B

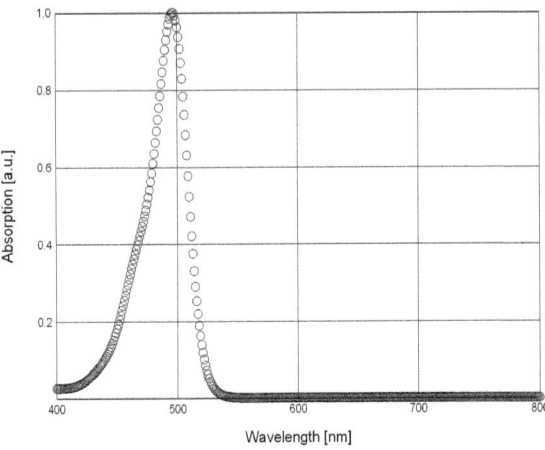

Figure 3.11: Absorption spectrum of Rhodamine 110

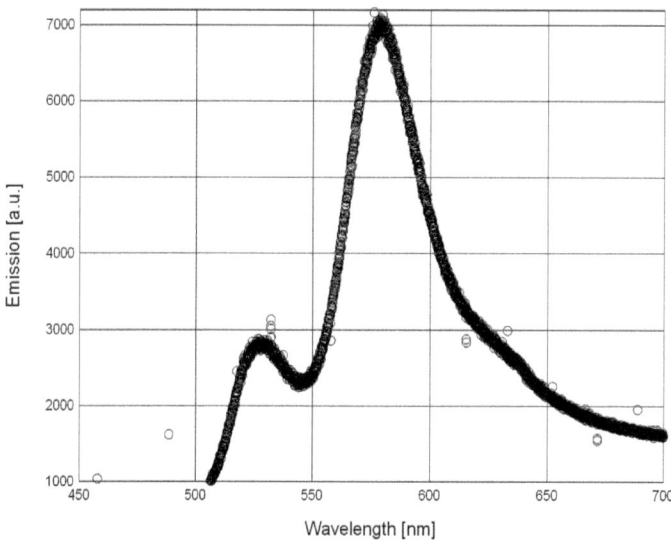

Figure 3.12: Emission spectrum of a mixture of Rhodamine B and Rhodamine 110 excited with a Nd:YAG laser ($\lambda_{ex} = 532$ nm)

Table 3.1: Properties of Rhodamine B and Rhodamine 110 in water at $T = 20\,°C$.

Dye	Molecular weight [kg/kmol]	λ_{abs} [nm]	λ_{em} [nm]
Rh B	479.02	554	575
Rh 110	366.80	496	520

this aspect measurements are performed in a Rayleigh–Bénard convection cell described in detail by Günther (2001b). 150 µg/l Rhodamine B were diluted in the water-filled convection cell and both walls were kept at constant temperatures ranging from 20 °C to 50 °C. To ensure a homogeneous temperature distribution the fluid was mixed with a magnetic stirrer between the measurements. A Nd:YAG laser was used as excitation light source. Figure 3.13 depicts the relation between the temperature and the emitted light intensity of Rhodamine B. These results exhibit a linear dependency of the emitted light intensity on the fluid temperature. Thus the choice of a linear calibration function for data analysis is justified.

3.3.4 Measurement Accuracy

Uncertainties in the PLIF measurements originate from the calibration and the measurement itself. During the calibration process the fluid temperature of the measurement plane inside the test section of the channel was recorded with a thermocouple having a sensitivity of 0.1 °C. For each temperature the calibration is performed with an image averaged over 10 recordings.

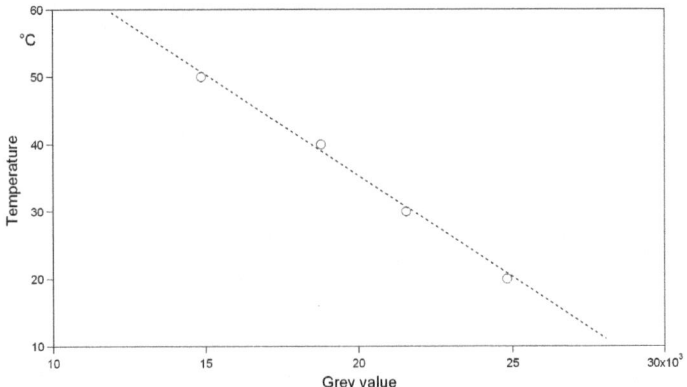

Figure 3.13: Relation between the temperature and the emitted light intensity of Rhodamine B.

The concentration of the tracer dyes and the optical path is kept constant between calibration and measurement, thus these effects are considered in the calibration function. We estimate the uncertainty in our temperature measurements to be on the order of 1 °C, which corresponds to approximately 4% of the bulk temperature.

3.4 Simultaneous Velocity and Scalar Measurements

To address the structure of turbulent scalar fluxes simultaneous measurements of the velocity and the scalar field are required. In the next sections the experimental setup and the experimental procedure to combine PIV with one–color LIF, respectively PIV with two–color LIF, is described.

3.4.1 Simultaneous Velocity and Concentration Measurements

The setup for simultaneous velocity and concentration measurements in the (x,y)–plane is depicted in Figure 3.14. A flashlamp–pumped dual Nd:YAG-laser provides the pulse light source for both PIV and LIF. For PIV the flow is seeded with hollow glass spheres with a mean diameter of 10 μm (density: 1.03 g/cm^3), the scattered light is recorded with a 10–bit CCD camera with a pixel-resolution of 1280 × 1024 pixels^2. As tracer for LIF Rhodamine B in aqueous solution with an inlet concentration of $1 \cdot 10^{-4}$ mol/l is used, the fluorescent light is recorded with a 12–bit CCD camera with a pixel-resolution of 1376 × 1040 pixels^2. To separate the Mie–scattering of the seeding particles from the fluorescent emission of the dye the PIV camera is equipped with a band–pass filter (passing light at 532 nm ± 2 nm), the PLIF camera with a high–pass filter (cutoff below 540 nm). To perform simultaneous measurements the cameras, the laser and the image acquisition software were synchronized by means of an adjustable timing unit (PIV Synchronizer, ILA GmbH).

We consider an ensemble of 1000 consecutive image pairs acquired at a frame rate of 4 Hz.

3.4 Simultaneous Velocity and Scalar Measurements

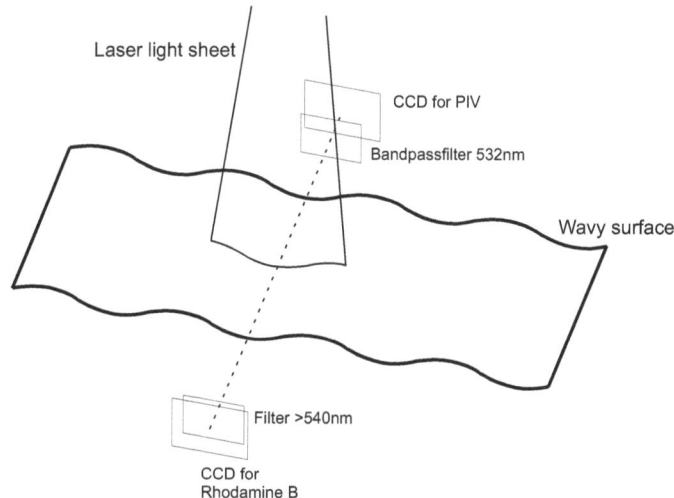

Figure 3.14: Sketch of the optical setup for simultaneous velocity and concentration measurements in the (x,y)–plane.

The post–processing of the PIV data is comprised of an adaptive cross–correlation algorithm (Scarano and Riethmuller (2000); Scarano (2002)), local filtering and interpolation of the filtered vectors. For LIF a single pixel calibration is employed to account for inhomogeneities in the laser light sheet. To account for any backscattering effects of the seeding particles on the emitted light the calibration is performed with the flow being seeded. In the post–processing the concentration information is then averaged over an area of 4×4 pixels^2 and is then mapped onto the velocity field.

3.4.2 Simultaneous Velocity and Temperature Measurements

The setup for simultaneous velocity and temperature measurements in the (x,y)–plane is depicted in Figure 3.15. Rhodamine B and Rhodamine 110, both at a concentration of 150 µg/l, are homogeneously distributed in the flow facility. The light sheet is provided by a Nd:YAG-laser, for PIV the flow is seeded with hollow glass spheres ($d_P = 10$ µm) whose scattered light is recorded with a 10–bit CCD camera (1280×1024 pixels^2). The PIV camera is equipped with a band–pass filter (532 nm ± 2 nm).

The spectral characteristic of the beamsplitter (Laseroptik GmbH) exhibits a high transmittance at 520 nm and a high reflectance at 575 nm. Thus the emitted light from Rhodamine B and Rhodamine 110 is separated between the two CCDs (1376×1040 pixels^2, 12–bit). Since the flow is seeded the Mie–scattering from the particles has to be removed from the emitted light of Rhodamine 110. This is accomplished by the filter mounted in front of the CCD for Rhodamine 110 which only transmits light with a wavelength smaller than 530 nm.

The temperature calibration is performed according to Sakakibara and Adrian (2004). We

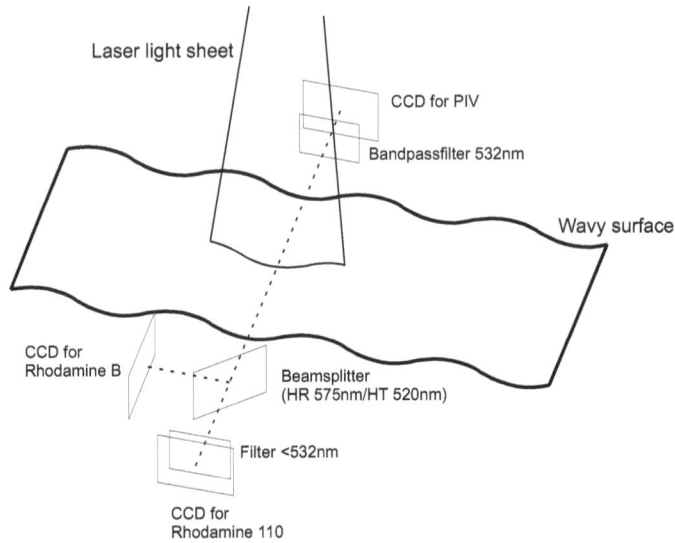

Figure 3.15: Sketch of the optical setup for simultaneous velocity and temperature measurements in the (x,y)–plane.

recorded two sets of reference images at uniform and known temperatures T_1 and T_2 and applied a linear interpolation to calculate the temperature

$$T = \frac{\gamma - \gamma_2}{\gamma_2 - \gamma_1}(T_2 - T_1) + T_2, \tag{3.15}$$

where γ_1 is the intensity ratio at temperature T_1, respectively γ_2 the intensity ratio at T_2. This calibration function is evaluated for each pixel. To account for any backscattering effects of the seeding particles the calibration is performed with the flow being seeded.

To perform simultaneous measurements the cameras, the laser and the image acquisition software were synchronized by means of an adjustable timing unit (PIV Synchronizer, ILA GmbH). We consider an ensemble of 1000 consecutive image pairs acquired at a frame rate of 4 Hz. The post–processing of the PIV data is comprised of an adaptive cross–correlation algorithm, local filtering and interpolation of the filtered vectors. The temperature information is averaged over an area of 8×8 pixels^2 and is then mapped onto the velocity field.

Simultaneous Transport of Mass and Momentum in Forced Convective Flows

Chapter 4

Influence of Wavy Surfaces on Coherent Structures in a Turbulent Flow

We describe how outer flow turbulence phenomena depend on the interaction with the wall. We investigate coherent structures in turbulent flows over different wavy surfaces and specify the influence of the different surface geometries on the coherent structures. The most important contribution to the turbulent momentum transport is attributed to these structures, therefore this flow configuration is of large engineering interest. In order to achieve a homogeneous and inhomogeneous reference flow situation two different types of surface geometries are considered: (i) three sinusoidal bottom wall profiles with different amplitude–to–wavelength ratios of $\alpha = 2a/\Lambda = 0.2$ ($\Lambda = 30$ mm), $\alpha = 0.2$ ($\Lambda = 15$ mm), and $\alpha = 0.1$ ($\Lambda = 30$ mm); and (ii) a profile consisting of two superimposed sinusoidal waves with $\alpha = 0.1$ ($\Lambda = 30$ mm). Measurements are carried out in a wide water channel facility (aspect ratio 12:1). Digital particle image velocimetry (PIV) is performed to examine the spatial variation of the streamwise, spanwise and wall-normal velocity components in three measurement planes. Measurements are performed at a Reynolds number of 22400, defined with the channel height H and the bulk velocity U_B. We apply the method of snapshots and perform a proper orthogonal decomposition (POD) of the streamwise, spanwise, and wall-normal velocity components to extract the most dominant flow structures. The structure of the most dominant eigenmode is related to counter–rotating, streamwise–oriented vortices. A qualitative comparison of the eigenfunctions for different sinusoidal wall profiles shows similar structures and comparable characteristic spanwise scales $\Lambda_z = 1.5H$ in the spanwise direction for each mode. The scale is observed to be slightly smaller for $\alpha = 0.2$ ($\Lambda = 15$ mm) and slightly larger for $\alpha = 0.2$ ($\Lambda = 30$ mm). This scaling for the flow over the basic wave geometries indicates that the size of the largest structures is neither directly linked to the solid wave amplitude, nor to the wavelength. The characteristic spanwise scale of the dominant eigenmode for the developed flow over the surface consisting of two superimposed waves reduces to $0.85H$. However, a scale in the order of $1.3H$ is identified for the second mode. The eigenvalue spectra for the superimposed waves is much broader, more modes contribute to the energy–containing range. The turbulent flow with increased complexity of the bottom surface is characterized by an increased number of dominant large–scale structures with different spanwise scales.

4.1 Introduction

Technical and geophysical relevant flows are characterized by high Reynolds numbers and complex boundaries. The turbulent flow over these rough or structured surfaces is associated with increased transport of species (heat or mass) and momentum. Differently shaped wavy walls as boundaries of the flow resemble the wall complexity in a well–defined manner and therefore serve as test cases for the study of wall influence on turbulence. We investigate the turbulent flow between a flat top and a complex bottom wall and describe the influence of different surface geometries on large–scale coherent structures in the region of the flow above the wall shear layer. In order to achieve a homogeneous and inhomogeneous reference flow situation two different types of surface geometries are considered: (i) three sinusoidal bottom wall profiles with different amplitude-to-wavelength ratios; and (ii) a profile consisting of two superimposed sinusoidal waves.

The flow over a train of solid waves is connected to a developing shear layer, formed by the separation of the flow shortly after the wave crest, which extends over the whole wavelength (Cherukat et al. (1998)). For smooth walls flow-oriented vortical eddies have been associated with large Reynolds stresses and with the production of turbulence in the viscous region close to the wall (Brooke and Hanratty (1993)). Günther and Rudolf von Rohr (2003), Kruse et al. (2003), and Kruse et al. (2006) investigated the structure and dynamics of turbulent motions in the outer part of the wall shear layer in a developed turbulent flow over waves and identified flow–oriented large–scale structures.

The literature on the stability of a sheared flow over solid waves suggests the Görtler mechanism (Görtler (1940); Saric (1994)) or the Craik–Leibovich type-2 mechanism (Phillips and Wu (1994); Phillips et al. (1996)) to produce or catalyse spanwise–periodic longitudinal vortices. The Görtler mechanism is based on a boundary–layer instability induced by the local wall curvature which then forms steady, flow–oriented, counter–rotating vortices (Görtler vortices). In the works of Phillips and Wu (1994) and Phillips et al. (1996) longitudinal flow structures are explained by a Craik–Leibovich type-2 (CL–2) instability. This instability is caused by the periodicity of the bounding surface rather than its local curvature. The CL-2 instability is referred to as catalysing longitudinal flow structures.

We address the characteristic scales and eigenvalue spectra of large–scale structures in the turbulent flow over differently shaped complex surfaces by means of a proper orthogonal decomposition (POD) of the velocity field obtained by stereoscopic particle image velocimetry (PIV). Thus we describe the influence of different surface geometries on these large–scale coherent structures.

4.2 Results

We use the method of snapshots and perform a proper orthogonal decomposition (POD) of the measured velocity fields according to section 2.4.

4.2.1 Results in the (x,y)–plane

The measurements in the (x,y)–plane are performed in a field of view which covers the whole region between the complex surface and the flat top wall, extending at least over one wavelength in the streamwise direction (FOV 1.1H (streamwise) × 1.0H (vertical)). The spatial resolution of the PIV data in this plane of measurement is 0.014H (streamwise) × 0.015H (vertical). Thus it is possible to determine the location of the most dominant flow structures in the flow field. Figure 4.1 depicts the contour plot of the streamwise component of the first eigenfunction obtained by a proper orthogonal decomposition of $u/U_B(x,y,z = 0.5B,t)$ for all wavy profiles at a Reynolds number of $Re_H = 22400$. The most energetic flow structure is located in the vicinity

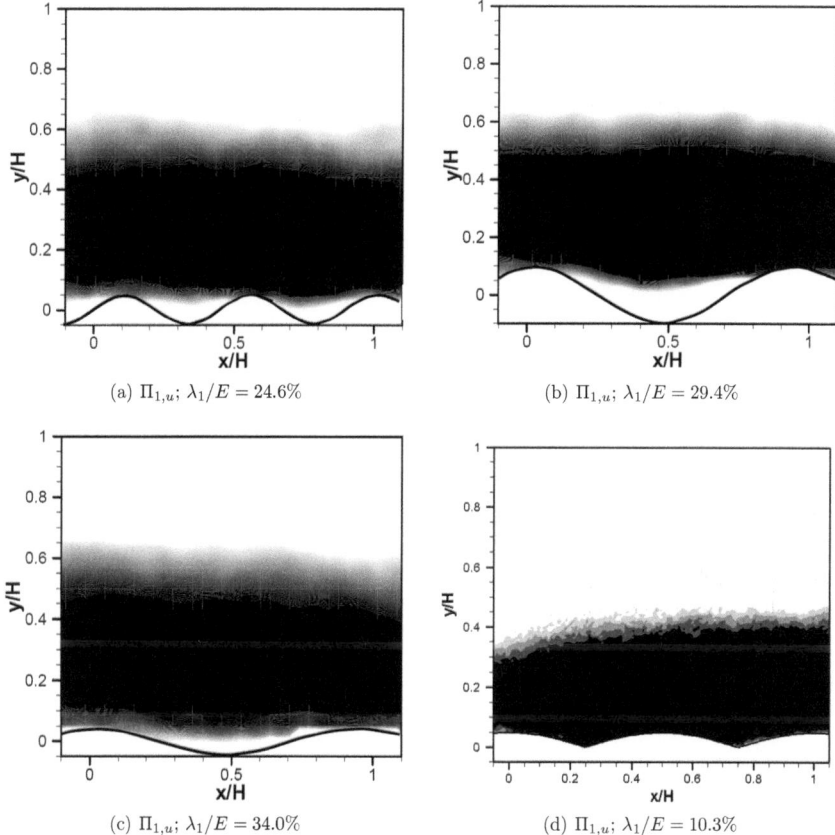

Figure 4.1: Comparison of the first eigenfunction for a decomposition of $u/U_B(x,y,z=0.5B,t)$ for (a) $\alpha = 0.2$ ($\Lambda = 15$ mm), (b) $\alpha = 0.2$ ($\Lambda = 30$ mm), (c) $\alpha = 0.1$ ($\Lambda = 30$ mm), and (d) superimposed waves, $\mathrm{Re}_H = 22400$.

of the complex surface. The maximum is found at a location of $y/H = 0.26$ for the profiles $\alpha = 0.2$ ($\Lambda = 15$ mm) and $\alpha = 0.1$ ($\Lambda = 30$ mm), for $\alpha = 0.2$ ($\Lambda = 30$ mm) slightly above and for the superimposed waves slightly below that value. We identify contributions to the kinetic energy of 34% for the profile $\alpha = 0.1$ ($\Lambda = 30$ mm), which decrease with decreasing wavelength to 24.6% for the profile $\alpha = 0.2$ ($\Lambda = 15$ mm), respectively with increasing amplitude to 29.4% for the profile $\alpha = 0.2$ ($\Lambda = 30$ mm). The lowest value of 10.3% is found for the profile consisting of the superimposed waves. Figure 4.2 depicts the fractional and cumulative kinetic energy contribution of the streamwise eigenvalues. For the flow over the basic wavy profiles more than 50% of the total energy is captured by less than 10 modes. With increasing wall complexity the eigenvalue spectrum becomes broader and therefore the number of important modes in the flow increases.

4.2.2 Results in the (x,z)–plane

The measurements in the (x,z)–plane are performed in a field of view which extends over a length corresponding to at least one channel height in streamwise direction and at least 1.5 channel heights in spanwise direction. For the three basic wave profiles the field of view is $2.2H$ (streamwise) $\times$ $3.0H$ (spanwise) with a resulting spatial resolution of $0.036H$ (streamwise) $\times$ $0.038H$ (spanwise), for the superimposed waves a smaller field of view of $1.1H$ (streamwise) $\times$ $1.5H$ (spanwise) with a resulting higher resolution of $0.018H$ (streamwise) $\times$ $0.019H$ (spanwise) is chosen. Thus we are able to address the spanwise scaling of the large–scale structures. Figs. 4.3 and 4.4 depict the contour plots of the streamwise component of the first two eigenfunctions obtained by a decomposition of $u/U_B(x,y/H = 0.26,z,t)$ for the four wall profiles. In this figure we marked the wave crest (continuous line) and the wave trough (dashed line) for the basic wave profiles. A qualitative comparison of the eigenfunctions for the three sinusoidal profiles exhibits similar structures and a characteristic spanwise scale in the order of $\Lambda_z = 1.5H$. This spanwise scale is also confirmed by the second eigenfunction. The highest energy contribution is found for the first mode of the profile $\alpha = 0.1$ ($\Lambda = 30$ mm). Similar to the results in the (x,y)–plane the energy contribution reduces with decreasing wavelength ($\alpha = 0.2$ ($\Lambda = 15$ mm)), respectively increasing amplitude ($\alpha = 0.2$ ($\Lambda = 30$ mm)). For the profile of the superimposed waves the energy contribution of the first mode is 10.4%. This value is larger than the one for the surfaces with an amplitude–to–wavelength ratio of $\alpha = 0.2$, but smaller than for the profile $\alpha = 0.1$ ($\Lambda = 30$ mm). The spanwise scale of the first eigenfunction for the superimposed waves is in the order of $0.85H$. However, this scaling is not confirmed by the second mode, which exhibits a larger scale in the order of $1.3H$, comparable to the spanwise distance of the first and second eigenmodes of the basic wave profiles. This supports the notion that by increasing the surface complexity the energy spectrum increases and thus the number of important modes exhibiting different scales is enlarged. This is confirmed in Figure 4.5 where the fractional and cumulative kinetic energy contribution of the streamwise eigenvalues for the decomposition $u/U_B(x,y/H = 0.26,z,t)$ is depicted. The eigenvalue spectra of the basic sinusoidal profiles becomes broader with decreasing wavelength ($\alpha = 0.2$ ($\Lambda = 15$ mm)), respectively increasing amplitude ($\alpha = 0.2$ ($\Lambda = 30$ mm)). For the surface described by the superposition of two sinusoidal waves the broadest spectrum is found, although not so pronounced as the spectra in the (x,y)–plane (Figure 4.2).

To quantitatively address the spanwise scaling of the large–scale structures we perform a streamwise averaging of the first two eigenfunctions as depicted in Figure 4.6. To compare the scaling between the different basic wave profiles the curves are shifted in the homogeneous

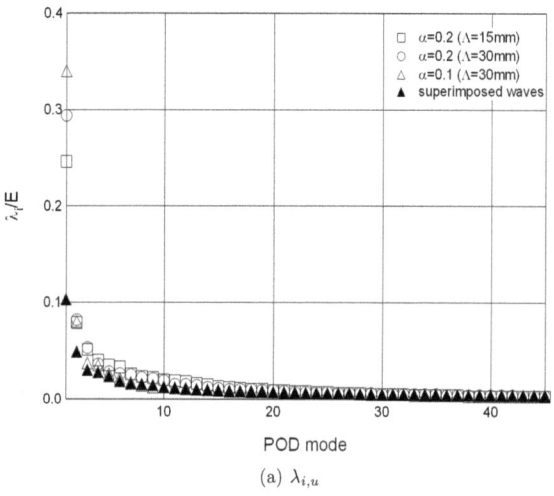

(a) $\lambda_{i,u}$

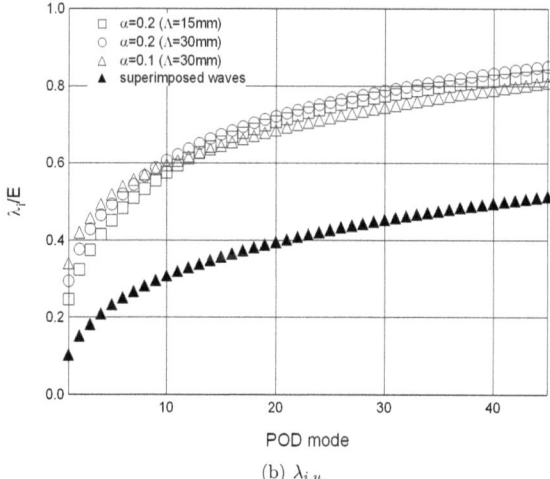

(b) $\lambda_{i,u}$

Figure 4.2: (a) Fractional and (b) cumulative kinetic energy contribution from streamwise eigenvalues for a decomposition of $u/U_B(x,y,z=0.5B,t)$, $\text{Re}_H = 22400$.

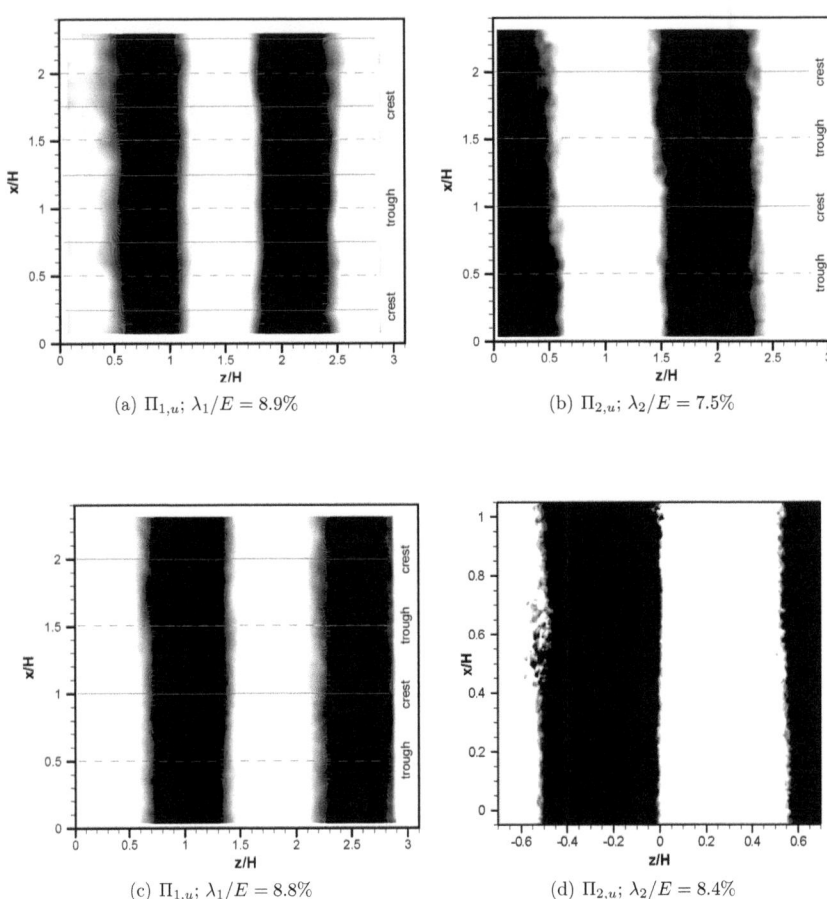

Figure 4.3: Comparison of the first eigenfunction for a decomposition of $u/U_B(x, y/H = 0.26, z, t)$ for (a) $\alpha = 0.2$ ($\Lambda = 15$ mm), (b) $\alpha = 0.2$ ($\Lambda = 30$ mm), (c) $\alpha = 0.1$ ($\Lambda = 30$ mm), and (d) for the superimposed waves, $Re_H = 22400$.

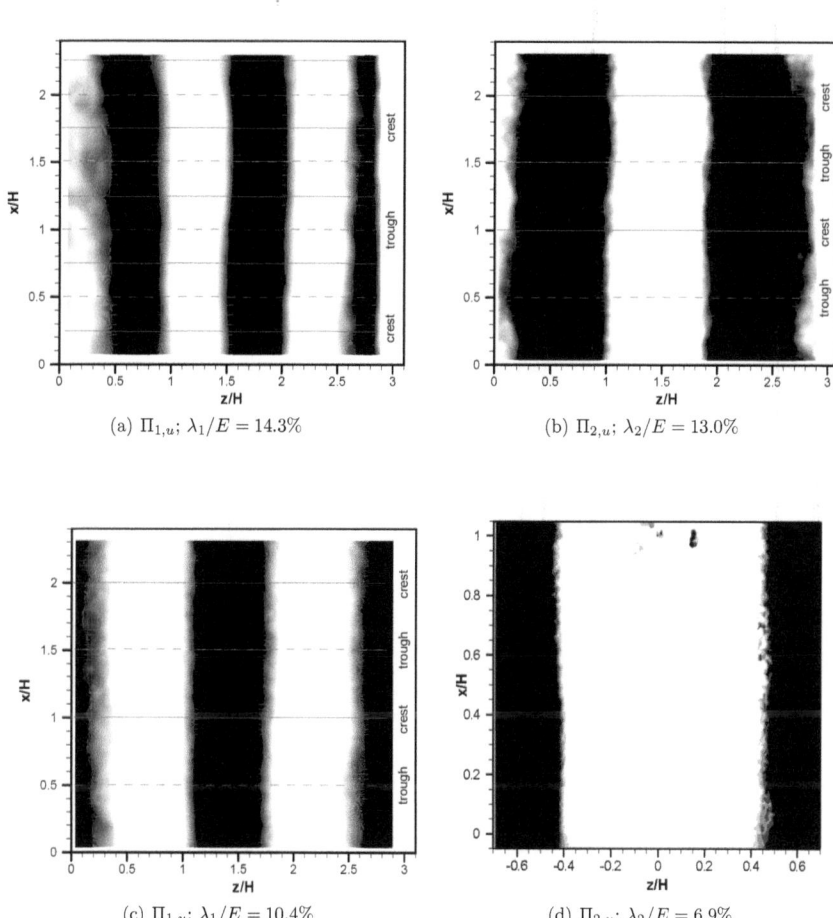

Figure 4.4: Comparison of the second eigenfunction for a decomposition of $u/U_B(x,y/H = 0.26,z,t)$ for (a) $\alpha = 0.2$ ($\Lambda = 15$ mm), (b) $\alpha = 0.2$ ($\Lambda = 30$ mm), (c) $\alpha = 0.1$ ($\Lambda = 30$ mm), and (d) for the superimposed waves, $Re_H = 22400$.

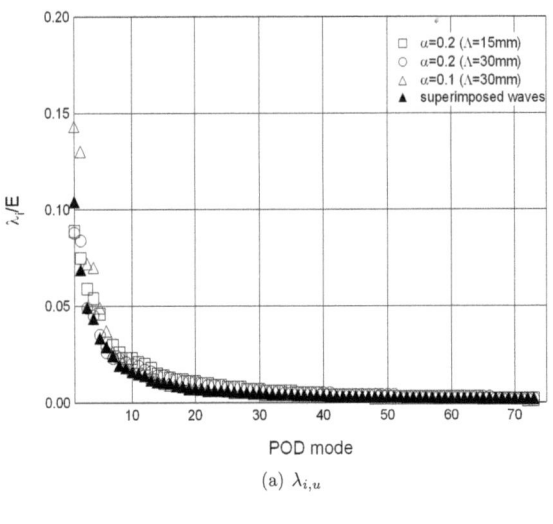

(a) $\lambda_{i,u}$

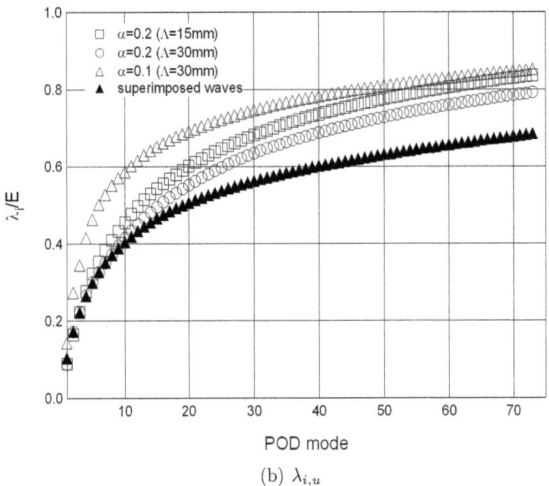

(b) $\lambda_{i,u}$

Figure 4.5: (a) Fractional and (b) cumulative kinetic energy contribution from streamwise eigenvalues for a decomposition of $u/U_B(x,y/H = 0.26,z,t)$, $\mathrm{Re}_H = 22400$.

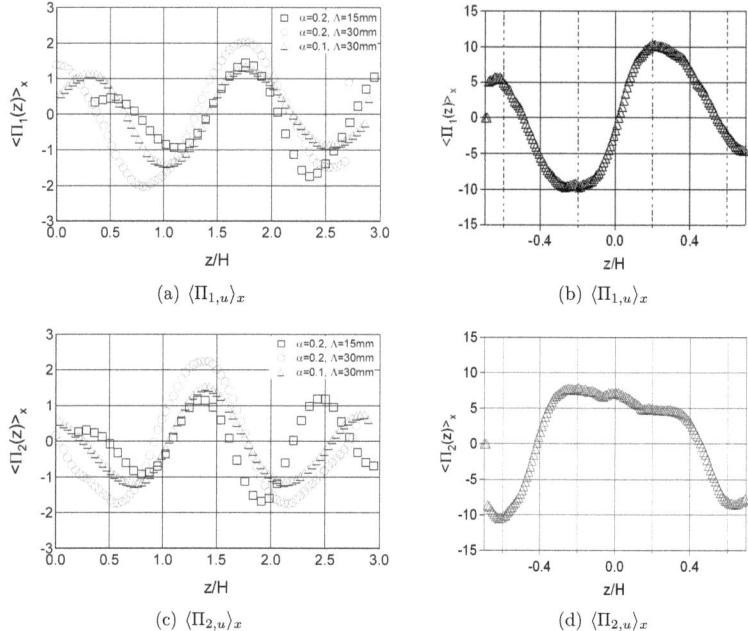

Figure 4.6: Comparison of streamwise–averaged eigenfunctions $\langle \Pi_{1,u} \rangle_x$ and $\langle \Pi_{2,u} \rangle_x$ for a decomposition of $u/U_B(x,y/H = 0.26,z,t)$ for (a) and (c) the wavy wall profiles and (b) and (d) the superimposed waves, $\mathrm{Re}_H = 22400$.

spanwise direction, so that one maximum coincides. We detect a characteristic spanwise scale of the first eigenfunction of $\Lambda_z = 1.5H$ for the profile $\alpha = 0.1$ ($\Lambda = 30$ mm). For the profile $\alpha = 0.2$ ($\Lambda = 30$ mm) with doubled amplitude the scaling increases to $\Lambda_z = 1.7H$. A spanwise distance of $\Lambda_z = 1.3H$ is found for the profile $\alpha = 0.2$ ($\Lambda = 15$ mm) with same amplitude but half the wavelength compared to $\alpha = 0.1$ ($\Lambda = 30$ mm). These scalings are also confirmed by the second eigenmode depicted in Figure 4.6(c). Thus we identify a characteristic spanwise scale of the first two eigenfunctions which is almost independent of the wall geometry. The scaling of the large–scale structures depends neither solely on the amplitude nor the wavelength of the sinusoidal surface.

For the surface consisting of two superimposed sinusoidal waves the spanwise scale of the first eigenmode of $\Lambda_z = 0.85H$ is confirmed by the averaging process. But for the second eigenfunction a larger spanwise distance of $\Lambda_z = 1.3H$ is found. This scaling is comparable to the one of the basic sinusoidal profiles found for the first two dominant modes.

4.2.3 Results in the (y,z)–plane

The measurements in the (y,z)–plane are performed in a field of view which covers the whole region between the complex surface and the flat top wall and extends at least over a length corresponding to one channel height in spanwise direction. For the basic wave profiles the field of view is $1.0H$ (vertical) × $1.6H$ (spanwise) with a resulting spatial resolution of $0.017H$ (vertical) × $0.02H$ (spanwise), for the superimposed waves a smaller field of view of $1.0H$ (vertical) × $1.0H$ (spanwise) with a resulting higher resolution of $0.012H$ (vertical) × $0.017H$ (spanwise) is chosen. The proper orthogonal decomposition is applied to the velocity vector field $\mathbf{U}/U_B(x/\Lambda = 0.5, y, z, t)$ to obtain structural information about the large–scale structures. Figure 4.7 depicts the first eigenfunction for the profile $\alpha = 0.2$ ($\Lambda = 15$ mm), exemplarily chosen for the basic wave profiles, and for the surface defined by the superposition of two sinusoidal profiles. The structure of the most dominant mode is that of counter–rotating vortices. Between these vortices fluid is transported towards and away from the surface. The centers of the vortices are found in the lower half of the channel at a wall–normal distance which corresponds to the maximum of the first eigenfunctions of the decomposition in the (x,y)–plane (Figure 4.1). The radius of the vortical fluid motion is slightly larger for the basic wave profiles, where a value of $0.8H$ is found. For the superimposed waves the radius decreases to $0.4H$. The spanwise scale of identical fluid movement (i.e. towards or away from the surface) is equal to the spanwise scale of the dominant eigenmode found in the (x,z)–plane. Therefore we identify the large–scale structures present in the vicinity of the complex surface as streamwise–oriented, counter-rotating vortices exhibiting a characteristic spanwise scale.

For all surfaces we found that eigenfunctions corresponding to higher POD modes and thus smaller energy contribution show characteristic smaller spanwise scales that may likely be connected to the local wall curvature, i.e. the Görtler instabilities (Görtler (1940); Saric (1994)).

4.3 Conclusions

In this study we addressed the influence of different surface geometries on the coherent structures in a turbulent flow. We obtained large image sequences in three planes of measurement of a turbulent channel flow at a Reynolds number of $Re_H = 22400$. We applied the method of snapshots and performed a proper orthogonal decomposition of the velocity field to extract structural information about the coherent structures.

From this orthogonal decomposition of velocity fields we find similar large–scale structures in the vicinity of the complex surface. These large–scale structures are identified as streamwise–oriented, counter–rotating vortices exhibiting a characteristic length scale in the spanwise coordinate direction. We quantitatively describe this spanwise scale by applying the proper orthogonal decomposition onto the streamwise velocity component in the (x,z)–plane. We found similar spanwise scales in the order of $\Lambda_z = 1.5H$ for the basic wavy surfaces of different amplitude-to–wavelength ratios in the first two eigenmodes. This finding is consistent with earlier studies by Günther and Rudolf von Rohr (2003) and Kruse et al. (2006). In addition, the oscillation and the location of the first two streamwise–averaged eigenfunction extrema are nearly identical. This scaling indicates that the size of the largest structures neither depends solely on the solid wave amplitude, nor on the wavelength.

For the profile described by the superposition of two sinusoidal waves a spanwise scale of $\Lambda_z = 0.85H$ was identified for the first eigenmode. However, this scaling is not confirmed by the second eigenfunction where a spanwise distance of $\Lambda_z = 1.3H$ is observed. Additionally, the location of the extrema of the second eigenmode is shifted by a spanwise distance of $z/H = 0.4$.

The eigenvalue spectra from the orthogonal decomposition in the (x,y)– and (x,z)–plane become increasingly broader for the profile with doubled amplitude ($\alpha = 0.2$ ($\Lambda = 30$ mm)),

4.3 Conclusions

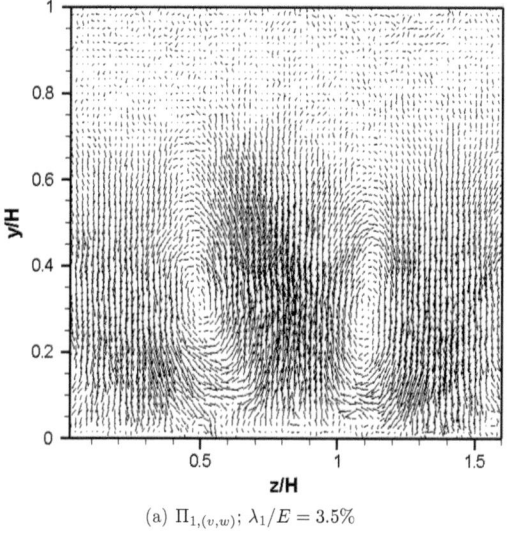

(a) $\Pi_{1,(v,w)}$; $\lambda_1/E = 3.5\%$

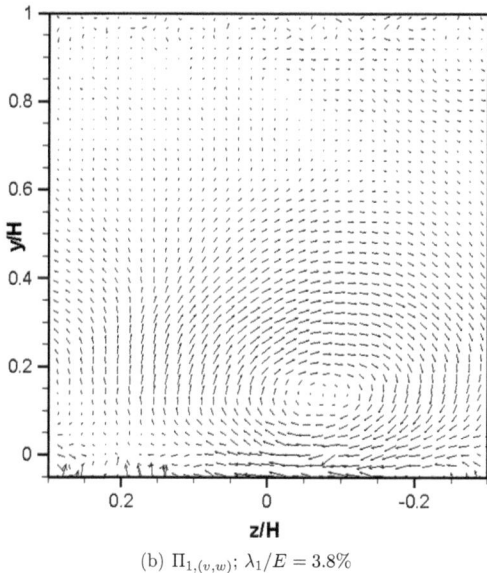

(b) $\Pi_{1,(v,w)}$; $\lambda_1/E = 3.8\%$

Figure 4.7: Comparison of the first eigenfunction for a decomposition of $\mathbf{U}/U_B(x/\Lambda = 0.5, y, z, t)$ for (a) $\alpha = 0.2$ ($\Lambda = 15$ mm) and (b) the superimposed waves, $\text{Re}_H = 22400$.

half the wavelength ($\alpha = 0.2$ ($\Lambda = 15$ mm)), and the superimposed waves. Thus by increasing the surface complexity more modes contribute to the energy containing range.

We conclude that by increasing the complexity of the bottom surface and thus altering the flow homogeneity from homogeneous in spanwise coordinate direction as for the basic wave profiles to completely inhomogeneous as for the superimposed waves the energy spectrum of the turbulent flow and the number of significant modes is increased. The flow over the superimposed waves can be described as the superposition of dominant eigenmodes with different spanwise scales.

Chapter 5
The Influence of Wavy Walls on the Transport of a Passive Scalar in Turbulent Flows

We carry out an experimental study to investigate the turbulent transport of a passive scalar in turbulent flows between a flat top wall and two different complex surfaces and describe the influence of the different surface geometries on the mixing properties. In order to achieve a homogeneous and inhomogeneous reference flow situation two different types of surface geometries are considered: a sinusoidal bottom wall profile, and a profile consisting of two superimposed sinusoidal waves. The laboratory investigations were performed by use of a combined digital particle image velocimetry (DPIV) and planar laser induced fluorescence (PLIF) technique to examine the spatial variation of the streamwise, spanwise and wall–normal velocity components, and to assess the concentration field of the scalar. Statistically sufficient image sequences (frame rate 4 Hz) are obtained from both DPIV and PLIF to calculate time–averaged statistics of the velocity field and the scalar field. We discuss the influence of the complex surfaces on large–scale structures present in the flow field and the transport of the passive scalar. A higher spanwise spreading of the scalar plume is observed compared to plane channel flows. Due to the influence of the complex surfaces a higher transport normal to the mean flow direction is found. The mixing properties of the turbulent flow are enhanced for the surface composed of two superimposed waves compared to the sinusoidal wavy surface. This is consistent with the scaling of large–scale structures found for this surface geometry.

5.1 Introduction

Relevant transport processes in technical applications as well as in nature are characterized by high Reynolds numbers, complex boundaries and the additional transport of a scalar (heat and mass). Through the interaction of the flow with complex surface coherent structures are formed which play an important role in turbulent transport processes and thus also have an impact on turbulent mixing processes and scalar transport. This experimental study addresses the influence of different complex boundaries on the transport properties of a passive scalar in a forced convective turbulent flow. Differently shaped wavy walls are chosen as boundaries of the flow to represent the wall complexity. In order to achieve a homogeneous and inhomogeneous reference flow situation two different types of surface geometries are considered: (i) a two–dimensional surface being homogeneous (i.e. translational invariant) in the spanwise direction described by a sinusoidal bottom wall profile propagating in the streamwise direction with an amplitude–to–wavelength ratio of $\alpha = 2a/\Lambda = 0.1$ ($\Lambda = 30$ mm); and (ii) a three–dimensional surface consisting of two superimposed sinusoidal waves with $\alpha = 0.1$ ($\Lambda = 30$ mm).

The investigation of turbulent mixing and passive scalars is an active area of research in which numerous research groups made their contributions in the past years. Turbulent mixing is the topic of a review by Dimotakis (2005). There three levels of mixing are defined: Level 1 being the mixing of passive scalars which do not couple back on the flow dynamics; Level 2 mixing is coupled to the dynamics of the flow, e.g. scalar transport in instability flows; Level 3 mixing is characterized by the transport of active scalar, e.g. buoyancy driven flows. According to this classification our study addresses Level 1 and Level 2 mixing. An additional overview about passive scalars in turbulent flows is given by Warhaft (2000). Hunt et al. (1979) and Snyder and Hunt (1984) investigated scalar transport in flows over complex terrain analytically and experimentally. The experiments were carried out in a towing tank, respectively in a wind tunnel, the surface complexity was represented by a fourth–order polynomial hill. In both cases the scalar was introduced through a point source. Their results indicated a dependence of the scalar dispersion on the source level height. Fackrell and Robins (1982) computed scalar statistics for elevated and ground level scalar sources in turbulent boundary layers. Their experiments were carried out in an open–circuit wind tunnel at two different source heights and combined measurements from a concentration probe (flame–ionization detection system) and hot wires to assess both the scalar and the velocity field. Their results showed that most of the production of fluctuations occurred in the vicinity of the source, which then decays in accordance with a balance between advection and diffusion. Differences in the vertical diffusion from sources at different levels and between near–field and far–field observations were related to the importance of the advection and diffusion terms. They also investigated the effects of the source size but could not detect any influences on the mean scalar field and the fluxes away from the immediate vicinity of the source. Crimaldi and Koseff (2001) employed two different techniques, namely planar LIF and a single–point laser–induced fluorescence probe, to assess the temporal and spatial structure of a turbulent plume. This study was then advanced to experimentally investigate the relationship between instantaneous scalar structure and the resulting mean scalar statistics of a scalar plume over a flat surface (Crimaldi et al. (2002)). This was accomplished by additionally adding a two–dimensional Laser–Doppler anemometer to the system to record vertical velocity profiles downstream of the source. Their results showed the largest structural variation in the vertical direction, the existence of a uniform layer of dye within the viscous sublayer, and above this layer stronger fluctuations. An additional overview over turbulent diffusion from point sources in complex flows is given in the review by Hunt (1985).

To allow the computation of second or higher order statistics of the velocity and the scalar (e.g. the turbulent scalar fluxes) a synchronization of the measurement system for the velocity and the scalar field is needed to acquire the measured signals simultaneously in space and time. Several papers report such simultaneous measurements, e.g. a pointwise measurement technique employed by Lemoine et al. (1996) by combining LIF and Laser Doppler velocimetry to investigate scalar transport in a turbulent jet; or a a whole–field measurement technique used by Law and Wang (2000) by means of a combination of digital particle image velocimetry (DPIV) and planar laser induced fluorescence (PLIF) to measure mixing processes in a turbulent jet. In the present study DPIV is performed to examine the spatial variation of the streamwise, spanwise and wall–normal velocity components in two orthogonal planes of measurement (Adrian (1991); Westerweel (1997); Raffel et al. (1998); Westerweel and van Oord (2000)). The scalar concentration is measured with PLIF (Karasso and Mungal (1997); Shan et al. (2004)). We combine these two techniques to simultaneously assess the velocity and concentration field and to compute the statistical properties of the flow. In addition we calculate the turbulent scalar fluxes to address the influence of wavy walls on transport processes.

The flow over a train of solid waves is connected to a developing shear layer, formed by the separation of the flow shortly behind the wave crest, which extends over the whole wavelength (Cherukat et al. (1998)). For smooth walls flow-oriented vortical eddies have been associated with large Reynolds stresses and with the production of turbulence in the viscous region close to the wall (Brooke and Hanratty (1993)). Günther and Rudolf von Rohr (2003), and Kruse et al. (2003, 2006) investigated the structure and dynamics of turbulent motions in a developed turbulent flow over waves and identified flow–oriented large–scale structures which contribute most to the momentum transport. In a recent study Kruse and Rudolf von Rohr (2006) investigated the transport of heat in a turbulent flow over a heated wavy wall. By employing a particle image thermometry technique the velocity and temperature fields were measured simultaneously. Quantitative agreement between large–scale thermal and momentum structures was found. We advance this study by addressing the effect of the bounding surface on the dispersion of a scalar emanating from a low momentum ground level point source in a forced convective turbulent flow over two different wavy walls by a combined DPIV/PLIF technique.

5.2 Flow Field

The measurements presented here are performed in two orthogonal planes, the (x,y)- and the (x,z)–plane, at a Reynolds number of $Re_H = 22400$, defined with the channel height H and the bulk velocity U_B.

5.2.1 Mean Velocity

In this section we discuss the influence of the surface geometry on the mean velocity profile in the (x,y)–plane. Therefore we plot wall–normal velocity profiles at distinct streamwise locations x/H along the complex surface. These locations are chosen as follows:

$x/H = 0.00$ Wave crest
$x/H = 0.25$ Inflection point of the wall profile
$x/H = 0.50$ Wave trough
$x/H = 0.75$ Inflection point of the wall profile
$x/H = 1.00$ Wave crest

Figure 5.1 depicts the mean velocity profile $\langle u \rangle/U_B$ for the flow over the sinusoidal surface. The influence of the structured surface is expressed by the asymmetric shape of the flow profile. The maximum value of the streamwise velocity component is found for a wall–normal distance of $y/H = 0.7$. Negative values of the velocity at the streamwise positions $x/H = 0.25$ and $x/H = 0.50$ indicate flow separation and a recirculation zone at the downstream side of the wave. Figure 5.2 depicts the mean velocity profile for the flow over the superimposed waves at the same streamwise locations. In contrast to the sinusoidal surface the outer flow is less affected by the superimposed waves. The velocity profile is nearly symmetric, the maximum value of the streamwise velocity component is found at a wall–normal position of $x/H = 0.55$. This means that the velocity profile is only shifted upward, since the height of the amplitude is $y/H = 0.05$. The effect of the surface geometry is visible in the near bottom wall region below $y/H < 0.1$. Due to the developing shear layer and the resulting higher shear compared to a flat wall the velocity profile develops a greater gradient. Because of the profile of the surface

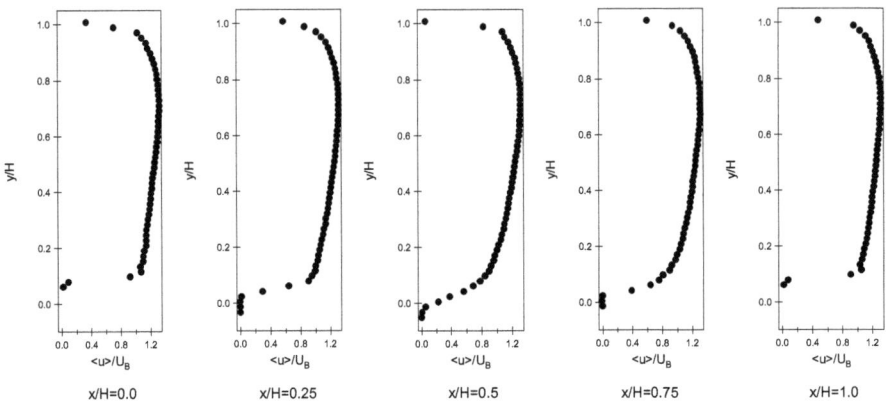

Figure 5.1: Mean velocity profile for the turbulent flow over the sinusoidal surface along one wavelength, $Re_H = 22400$.

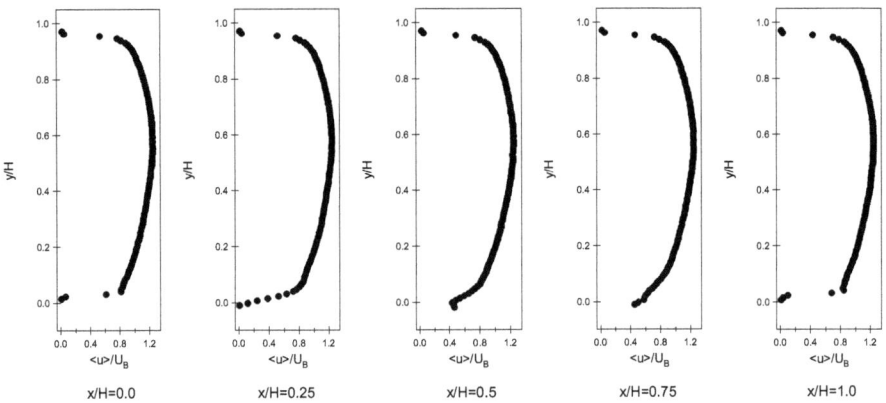

Figure 5.2: Mean velocity profile for the turbulent flow over the superimposed waves, $Re_H = 22400$.

and the resulting masking of the wave trough in this plane of measurement we are not able to confirm the existence of a recirculation zone.

5.2.2 Shear Layer

To estimate the influence of the complex surfaces on turbulence statistics we calculate the Reynolds shear stress. The flow over a wavy wall is associated with a shear layer developing behind the wave crest and extending over the whole wavelength (Hudson et al. (1996)). Figure 5.3 depicts the contour plots of the Reynolds shear stress in the (x,y)-plane for both surface geometries. The location of the shear layer can be identified in both cases. For the flow over the superimposed waves the shear layer is located closer to the complex surface, in a region between $-0.03 < y/H < 0.08$. The corresponding region for the sinusoidal profile is $0.04 < y/H < 0.15$. This result is in agreement with the shape of the profiles of the mean velocity. For the superimposed waves the region of highest shear is found in a region of $y/H < 0.1$ which corresponds to the large gradient in the mean velocity profile in this region. This means that compared to the sinusoidal wave the presence of the superimposed waves affects a smaller part of the flow field.

5.3 Scalar Transport

In this section we address the transport of a passive scalar. Therefore we apply Reynolds averaging to decompose the velocity and the concentration field into a mean ($\langle \cdot \rangle$) and fluctuating ($\cdot '$) part.

$$u_i = \langle u_i \rangle + u_i' \tag{5.1}$$

$$c = \langle c \rangle + c' \tag{5.2}$$

Thus we compute the mean concentration profiles and the root mean square (rms) of the concentration fluctuations. In addition we calculate the scalar fluxes $\langle u_i' c' \rangle$ which are important terms to close the scalar conservation equation (e.g. Pope (2000))

$$\frac{\overline{D}\langle c \rangle}{\overline{D}t} = \frac{\partial}{\partial x_i}\left(D\frac{\partial \langle c \rangle}{\partial x_i} - \langle u_i' c' \rangle\right) \tag{5.3}$$

5.3.1 Concentration Field

Results in the (x,y)-plane

The measurements in the (x,y)-plane are performed in a field of view which covers the whole region between the complex surface and the flat top wall, extending at least over one wavelength in streamwise direction. Thus it is possible to investigate the development of the scalar plume emanating from the point source located at $x/H = 0.00$, $y/H = 0.05$. Figure 5.4 depicts the contour plot of the mean concentration field and the root mean square of the concentration fluctuations for the turbulent flow over the sinusoidal surface. All values are made dimensionless with the dye concentration c_0 at the outlet of the point source. The scalar is convected with the mean flow along the wavy surface and continuously disintegrates. In the mean concentration profile no transport in wall-normal direction is observed. However, in the

5 Transport of a Passive Scalar in Turbulent Flows

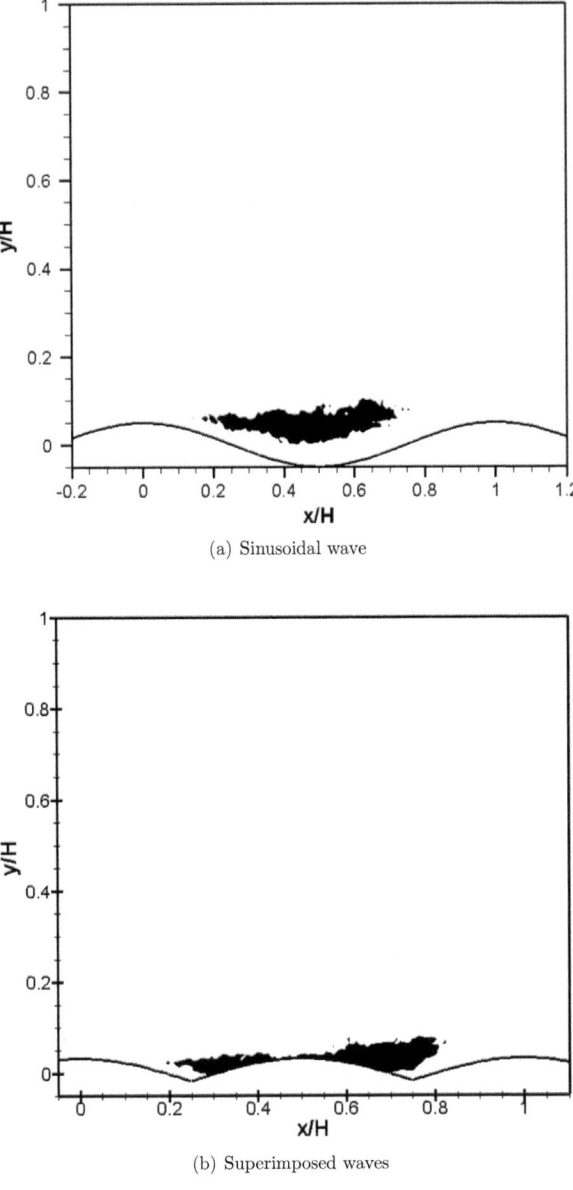

(a) Sinusoidal wave

(b) Superimposed waves

Figure 5.3: Contour plot of the Reynolds shear stress in the (x,y)–plane to locate the shear layer (threshold $\langle u'v' \rangle / U_B < -0.01$). Flow direction is from left to right, $\mathrm{Re}_H = 22400$.

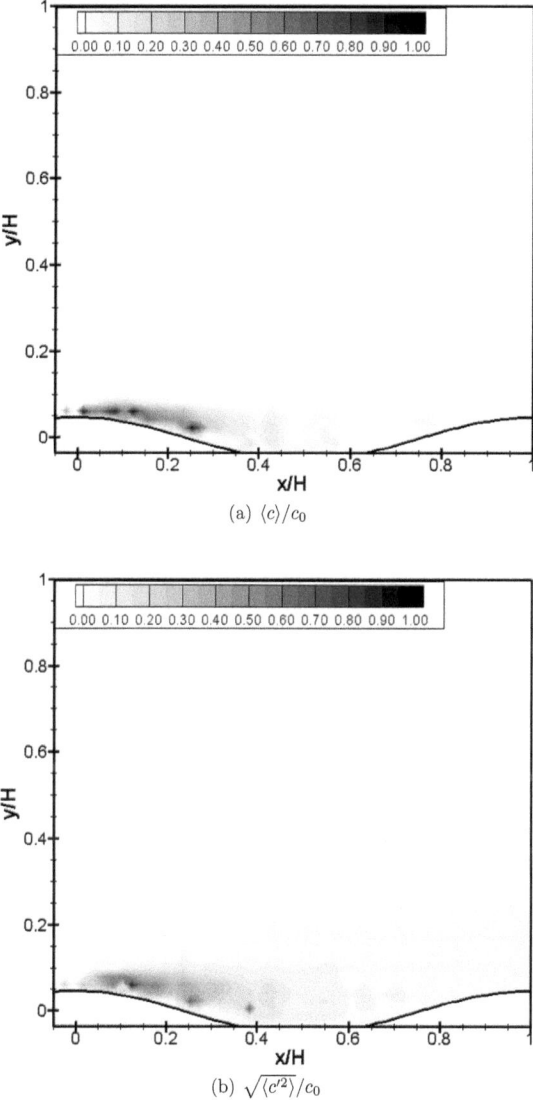

Figure 5.4: Contour plots of the mean concentration field and the root mean square of the concentration fluctuations for the turbulent flow over the sinusoidal surface. The point source is located at $x/H = 0.00, y/H = 0.05$, $\mathrm{Re}_H = 22400$.

instantaneous realizations rare bursting events occur and thus have negligible impact on the mean. Figure 5.5 depicts the contour plot of the mean concentration field and the root mean square of the concentration fluctuations for the turbulent flow over the superimposed waves. As already observed for the sinusoidal wave the scalar is transported along the complex surface and disintegrates. Contrary to the results found for the sinusoidal surface the root mean square of the concentration fluctuations exhibits higher values in the wall–normal direction for the streamwise positions $x/H > 0.3$. This supports the notion that the superimposed waves promote the scalar transport in the vertical direction.

Results in the (x,z)–plane

The measurements in the (x,z)–plane are performed in a field of view which extends over a length corresponding to at least 0.5 channel heights in streamwise and spanwise direction, the point source is located at $x/H = 0.0$, $z/H = 0.0$. Thus we are able to address the spreading of the scalar plume. The laser light sheet is adjusted to illuminate a plane 1.5 mm above the wavy surface. Figure 5.6 depicts the contour plots of the mean concentration field and the root mean square of the concentration fluctuations for the turbulent flow over the sinusoidal surface.

The constant spreading of the scalar plume is observed which will be further characterized in section 5.3.2. The contours of the rms–values of the concentration depict the transport normal to the mean flow.

Figure 5.7 depicts the contour plots of the mean concentration field and the root mean square of the concentration fluctuations for the turbulent flow over the superimposed waves. Again the constant spreading of the scalar plume is observed and in the rms–values the transport of the scalar normal to the mean flow direction is found. A decrease of the mean concentration is evident for $x/H > 0.3$, the scalar is transported in vertical direction and is thus diluted in the plane of the laser light sheet. Thus the scalar transport is additionally enhanced in vertical direction compared to the sinusoidal wave.

5.3.2 Spreading of the Scalar Plume

To quantitatively compare the scalar plumes for both surfaces the normalized mean concentration is plotted at constant streamwise locations $x/H = 0.1$, $x/H = 0.2$, $x/H = 0.3$, $x/H = 0.4$, and $x/H = 0.5$. Figure 5.8 depicts the profiles of the mean concentration for the sinusoidal profile (left column) and the superimposed waves (right column).

To describe the spreading of the scalar plume the spreading rate S is calculated according to Pope (Pope (2000))

$$S = \frac{\mathrm{d}z_{1/2}}{\mathrm{d}x} \tag{5.4}$$

where $z_{1/2}$ is the half–width of the scalar plume. For the sinusoidal surface a value of $S = 0.35$ is found, which is significantly larger than for a free turbulent jet ($S = 0.1$). Thus the spanwise transport is enhanced due to the influence of the surface. The observed spreading rate for the sinusoidal surface agrees with the spreading rate of $S = 1/3$ of a jet in crossflow configuration with a velocity ratio (jet–to–crossflow) of 5.7 (Su and Mungal (2004)). Calculating the spreading rate of the scalar plume for the flow over the superimposed waves in the region $0.0 < x/H < 0.25$ yields $S = 0.45$. It can be observed that mean concentration profiles of the scalar plume are broader and show lower normalized concentration values compared to the scalar plume over the

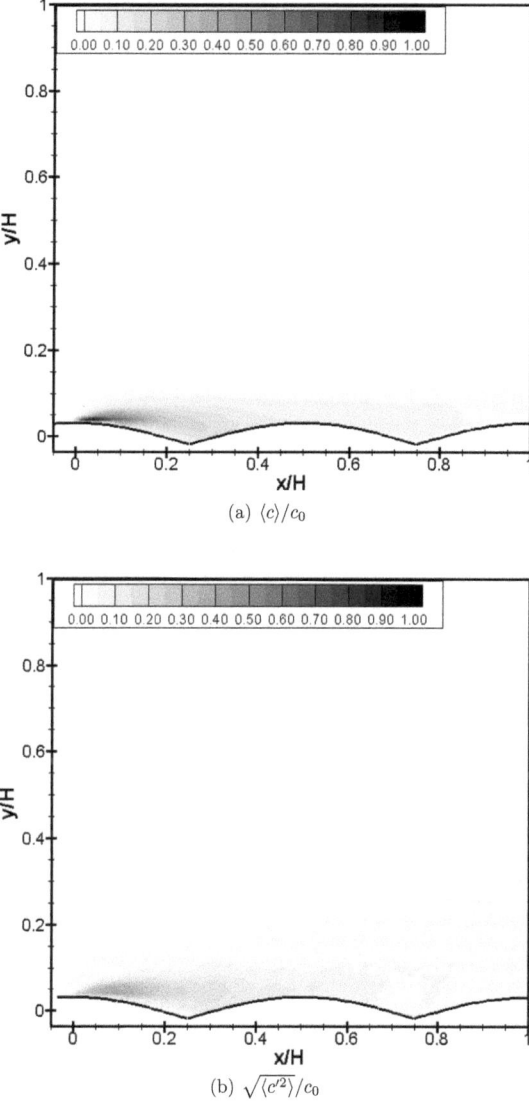

(a) $\langle c \rangle / c_0$

(b) $\sqrt{\langle c'^2 \rangle}/c_0$

Figure 5.5: Contour plots of the mean concentration field and the root mean square of the concentration fluctuations for the turbulent flow over the superimposed waves. The point source is located at $x/H = 0.00, y/H = 0.05$, $\mathrm{Re}_H = 22400$.

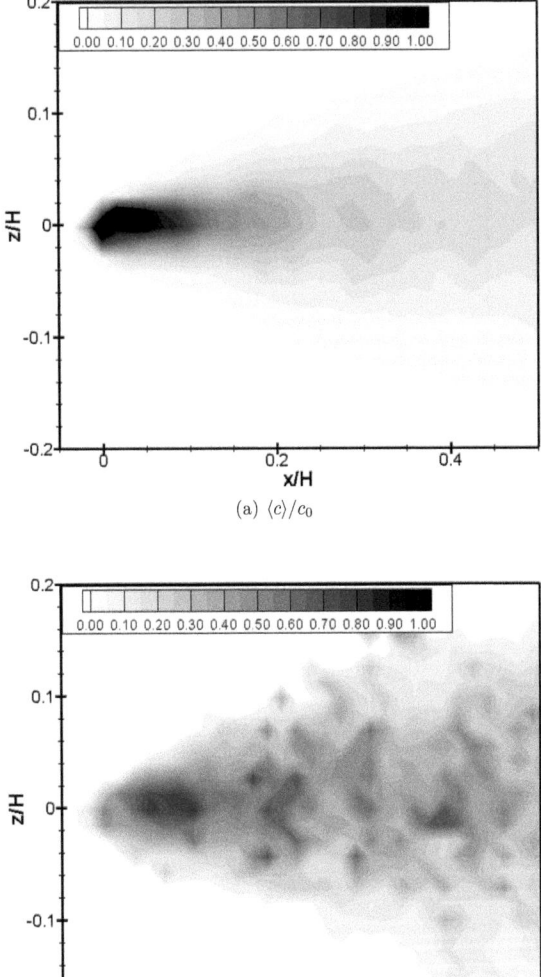

Figure 5.6: Contour plots of the mean concentration field and the root mean square of the concentration fluctuations for the turbulent flow over the sinusoidal surface. The point source is located at $x/H = 0.0, z/H = 0.0$, $\mathrm{Re}_H = 22400$.

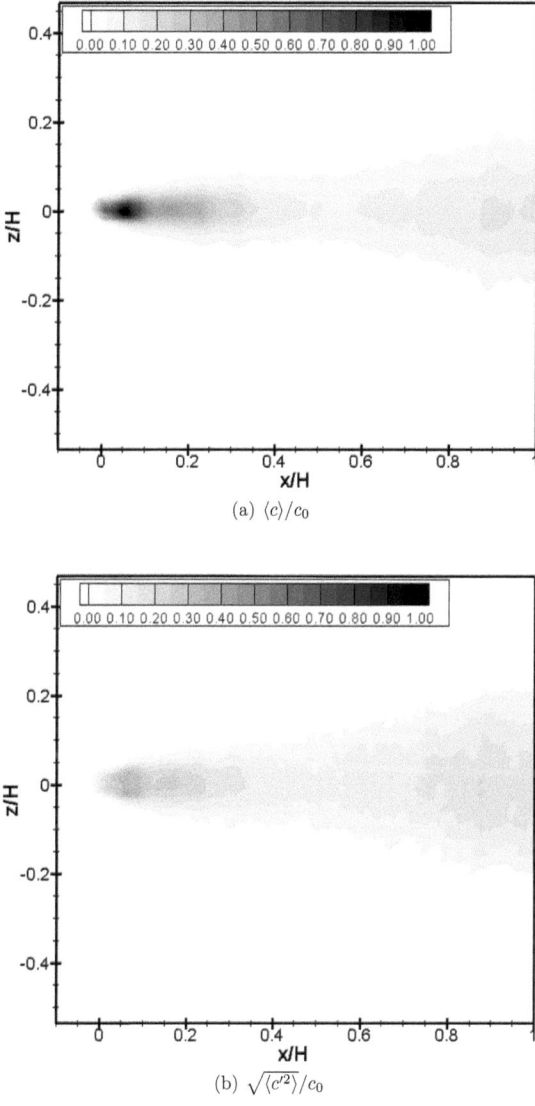

Figure 5.7: Contour plots of the mean concentration field and the root mean square of the concentration fluctuations for the turbulent flow over the superimposed waves. The point source is located at $x/H = 0.00, z/H = 0.0$, $\text{Re}_H = 22400$.

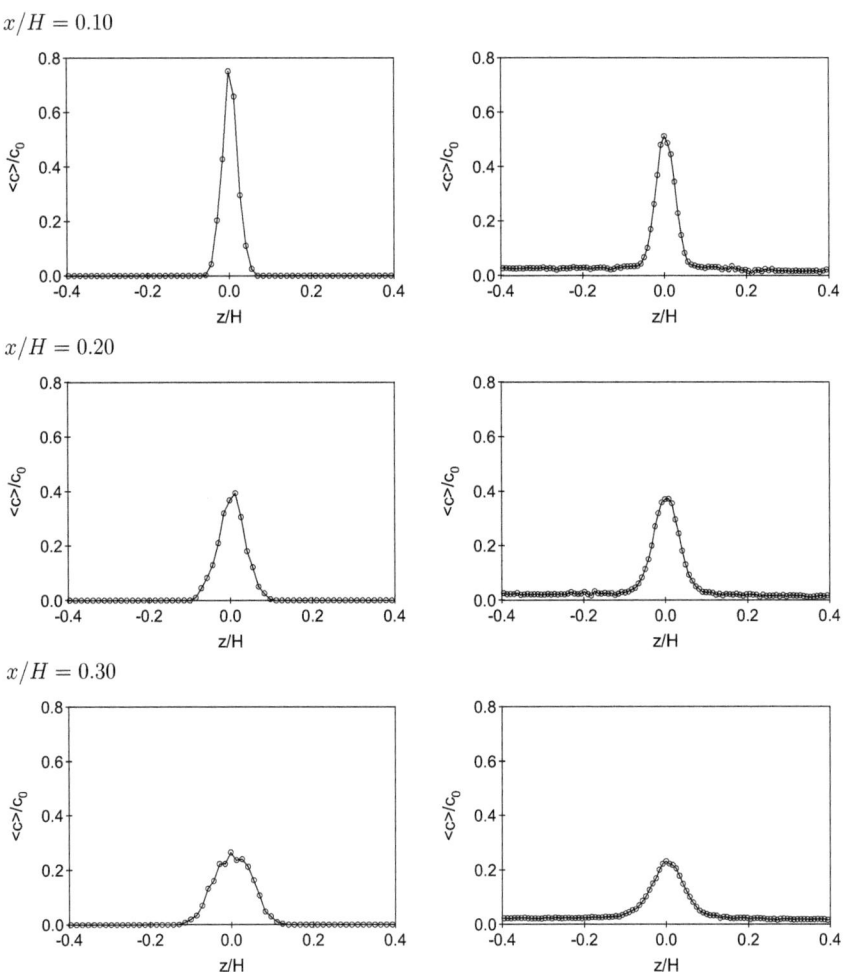

Figure 5.8: Profiles of the mean concentration $\langle c \rangle/c_0$ at constant streamwise locations x/H for the sinusoidal profile (left column) and the superimposed waves (right column), $\text{Re}_H = 22400$.

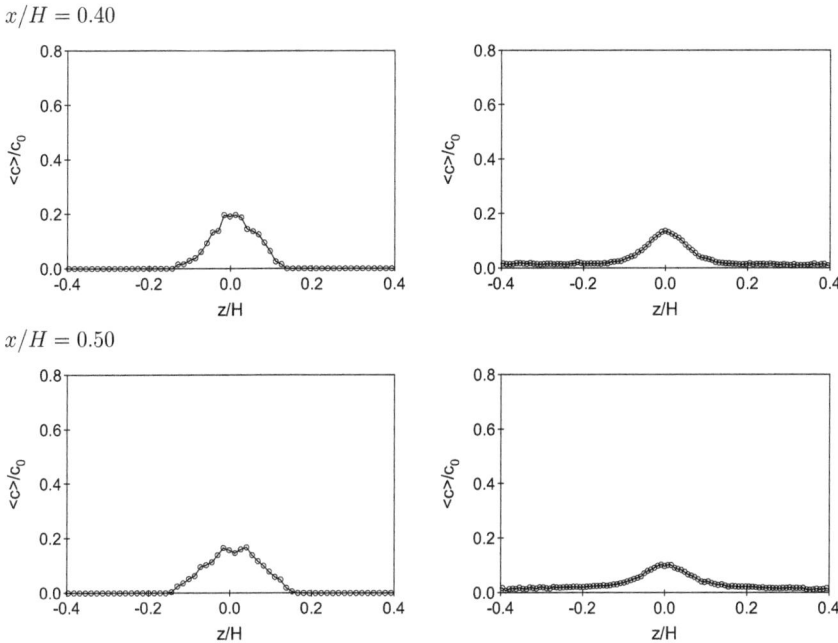

Figure 5.8: (Continued) Profiles of the mean concentration $\langle c \rangle / c_0$ at constant streamwise locations x/H for the sinusoidal profile (left column) and the superimposed waves (right column), $Re_H = 22400$.

sinusoidal surface. Thus the scalar transport is additionally enhanced in the spanwise direction compared to the sinusoidal wave.

5.3.3 Scalar Fluxes

To verify the findings so far we calculate the turbulent fluxes $\langle u'c' \rangle$ and $\langle v'c' \rangle$ in the (x,y)–plane. Figure 5.9 depicts the contour plots of the normalized turbulent scalar fluxes in streamwise and wall–normal direction for the flow over the sinusoidal surface. Normalization is accomplished with the bulk velocity U_B and the dye concentration c_0 at the outlet of the point source. Figure 5.10 depicts the contour plots of the normalized turbulent scalar fluxes for the superimposed waves. It is observed that the scalar transport normal to the mean flow is more pronounced for the flow over the superimposed waves. It can be concluded that structured surfaces enhance the mixing properties of a turbulent flow. This enhancement is larger for the flow situation which is inhomogeneous in all three coordinate directions.

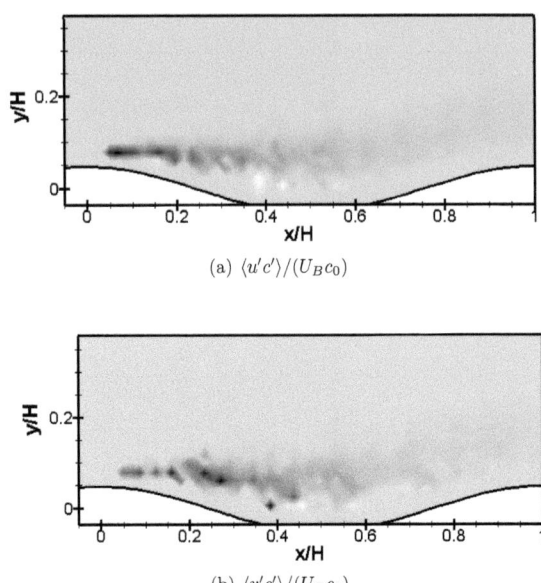

Figure 5.9: Contour plots of the turbulent scalar flux for the flow over the sinusoidal surface. The point source is located at $x/H = 0.00, y/H = 0.05$, $\text{Re}_H = 22400$.

5.4 Conclusions

We apply a combined DPIV/PLIF technique to investigate the effects of structured surfaces on the transport of a passive scalar in a turbulent flow. The structured surfaces are chosen in a way to resemble a homogeneous and an inhomogeneous reference flow situation. By recording long image time series the statistics of the velocity and the concentration field, as well as the turbulent scalar fluxes are calculated. Large–scale structures in the flow field are extracted by means of a proper orthogonal decomposition. Compared to a free turbulent jet the influence of the structured surfaces significantly enhances the scalar transport normal to the mean flow direction. The scalar plumes in the turbulent flow over the wavy surfaces exhibit larger values of the spreading rate compared to free jets. In addition there is no complete 'smoothing' of the instantaneous structures in the mean scalar field.

By comparing mean concentration profiles at constant streamwise locations along the complex surface it is observed that the surface composed of two superimposed sinusoidal waves promotes the scalar transport in spanwise and vertical direction compared to the basic sinusoidal surface. By calculating the streamwise and wall–normal turbulent scalar fluxes in the (x,y)–plane this result is confirmed. Due to the flow inhomogeneity in all three coordinate directions and the smaller spanwise scale of the large–scale structures the mixing properties of the turbulent flow are significantly enhanced.

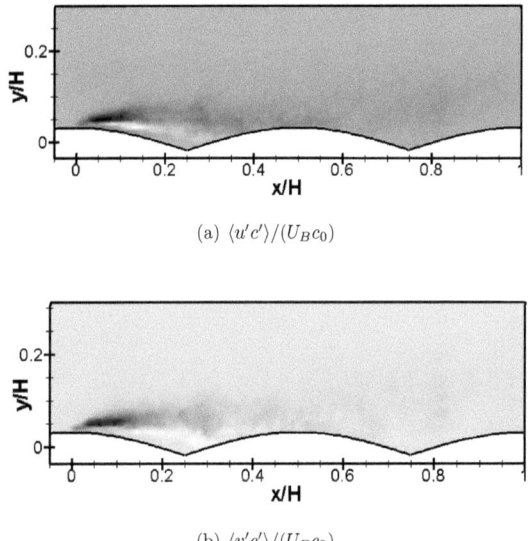

(a) $\langle u'c'\rangle/(U_B c_0)$

(b) $\langle v'c'\rangle/(U_B c_0)$

Figure 5.10: Contour plots of the turbulent scalar flux for the flow over the superimposed waves. The point source is located at $x/H = 0.00, y/H = 0.05$, $\mathrm{Re}_H = 22400$.

Simultaneous Transport of Mass, Heat and Momentum in Mixed Convective Flows

Chapter 6

Experimental Study of Heat Flux in Mixed Convective Flow over Solid Waves

In this experimental study we address transport processes in a mixed convective flow over a heated wavy surface. Therefore we combine digital particle image velocimetry (DPIV) and two–color planar laser induced fluorescence (PLIF) to simultaneously measure the velocity and temperature field. We propose to use the dye combination of Rhodamine B and Rhodamine 110, both excited with the Nd:YAG laser also used for the PIV measurements. We investigate the influence of mixed convection over a wavy surface on the velocity field, turbulence statistics, the temperature field and the heat flux. By computing these quantities we find a correlation between the maximum in the Reynolds stress profiles and the magnitude of the heat flux vector, thus regions of maximum momentum and scalar transport coincide. In addition we apply a proper orthogonal decomposition (POD) to extract the most dominant flow structures in a measurement plane above the wavy surface. This first POD mode is identified as streamwise–oriented, counter–rotating vortices whose spanwise scaling is also correlated with the magnitude of the heat flux.

6.1 Introduction

Flows in the mixed convective regime are present in relevant technical and geophysical flow situations. Especially in technical applications the flow boundaries resemble complex wall geometries, e.g. undulations in heat exchangers to enhance transport processes (e.g. Rush et al. (1999); Dellil et al. (2004)). Previous studies of mixed convective flows focused mainly on horizontal parallel plate configurations. Osborne and Incropera (1985a,b) and Maughan and Incropera (1989) investigated laminar, transitional, and turbulent mixed convection heat transfer for a horizontal parallel plate water channel experimentally. They found an increase in heat transfer due to buoyancy effects which is less pronounced for higher Reynolds numbers and is preceded by the onset of a secondary flow. The structure of this secondary flow in a mixed convective air flow through a horizontal plane channel was investigated numerically and experimentally by Yu, Chang, Huang and Lin (1997); Yu, Chang and Lin (1997). Their results showed that this secondary flow is in the form of longitudinal vortex rolls for high enough Reynolds numbers. To enhance the complexity of the investigated flow situation wavy walls are often chosen as a well–defined bounding surface of the flow. Several numerical studies addressed convective heat transfer in channel flow bound by one or two wavy walls (e.g. Dellil et al. (2004); Metwally and Manglik (2004)). In a recent experimental study Kruse and Rudolf von Rohr (2006) investigated the transport of heat (as a passive scalar) in a turbulent flow over a heated wavy wall. By employing a particle image thermometry technique the velocity and temperature fields were measured simultaneously. Quantitative agreement between large–scale thermal and momentum structures was found. The present work advances this study by addressing buoyancy effects induced by mixed convection from a wavy surface on transport processes by applying a combined digital particle image velocimetry and laser–induced fluorescence technique to simultaneously measure the velocity and the temperature field.

Non–invasive measurement techniques to assess the temperature field in a fluid is another important topic in recent research. One common technique is based on liquid crystals (Dabiri and Gharib (1991)) which is known as particle image thermometry and can also be used together with PIV to obtain simultaneous measurements (Kruse and Rudolf von Rohr (2006)). To assess the temperature field thermochromic liquid crystals dispersed in the fluid are illuminated with a sheet of white light. The color distribution reflected by the particles is recorded by a color CCD. In a first step the red, green, and blue intensities of the color distribution are converted into a local intensity, local hue, and saturation. This process is also known as RGB to HSI conversion. The temperature information is obtained by providing a calibration function between the local fluid temperature and the local hue. Major drawbacks of this technique are the need for a color CCD to record the color information emitted from the liquid crystals and, if simultaneous measurements are desired, the use of two different light sources which makes it complicate to synchronize the whole measurement system. Another whole–field technique is based on laser–induced fluorescence where temperature sensitive fluorescent dyes excitable by laser light are employed. As light sources, both constant wavelength Ar^+ ion lasers and pulsed Nd:YAG lasers are possible. However, to examine larger regions in the flow field and to acquire instantaneous images the use of pulsed Nd:YAG lasers is preferable (Karasso and Mungal (1997)). Coolen et al. (1999) report on one–color LIF temperature measurements with Rhodamine B as temperature sensitive dye excited with a Nd:YAG laser. The variation of the excitation light intensity, caused by different effects such as optical imperfections or even refraction of the light passing through the thermal field itself, is a source of measurement inaccuracy. Thus Sakakibara and Adrian (1999, 2004) proposed the use of two–color LIF where one fluorescent dye is temperature sensitive, the other insensitive and used as correction for inhomogeneities in the laser light sheet. The ratio of the emitted light of both dyes is then calibrated and used to calculate the temperature in the flow field. By applying this method of two–color LIF the concentration ratio of the two dyes has to be known and kept constant during measurement, which might pose a difficulty for example in turbulent diffusion processes. To address this problem the two–color/single–dye technique was introduced. Bruchhausen et al. (2005) report on two–color LIF with the use of two spectral bands of Rhodamine B as single fluorescent dye. Therefore these spectral bands need to exhibit a strong difference in their temperature sensitivity. By applying this technique, the ratio of the emitted fluorescence signal depends solely on the temperature of the fluid.

This experimental study addresses the mixed convective flow between a flat top wall and a heated wavy bottom wall. Thus transport processes in a complex flow situation, characterized by buoyancy effects, separation and reattachment of the flow due to the presence of the wavy wall, are evaluated. We present simultaneous measurements of the two–dimensional fluid velocity and temperature field by using a combined digital particle image velocimetry (DPIV) and two–color planar laser–induced fluorescence (PLIF) technique.

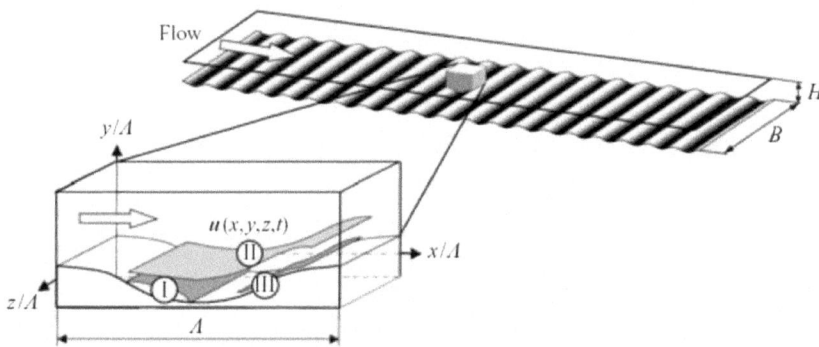

Figure 6.1: Schematic of (I) the separation region, and the regions (II) of maximum positive and (III) maximum negative Reynolds shear stress for a flow situation with separation.

6.2 Flow Description

We investigate the mixed convective channel flow between a heated sinusoidal bottom surface and a flat top wall. The mixed convective flow is characterized by the Reynolds number and the Grashof number. The Reynolds number Re_H is calculated according to

$$\text{Re}_H = \frac{U_B H}{\nu}, \tag{6.1}$$

where ν denotes the kinematic viscosity, H is the total height of the channel, and U_B the bulk velocity. The Grashof number Gr_H is given by

$$\text{Gr}_H = \frac{g H^3 \Delta T \beta}{\nu^2}, \tag{6.2}$$

where ΔT denotes the temperature difference between the bottom and top wall, and β is the volumetric thermal expansion coefficient. The ratio between the Grashof and the Reynolds number squared ($\text{Gr}_H/\text{Re}_H^2$) characterizes the convective regime of the flow. This ratio is in the range of unity for mixed convection, much smaller than one for forced convection, and much greater than one for natural convection (Incropera and DeWitt (2002)).

The isothermal flow over a train of solid waves is connected to a developing shear layer, formed by the separation of the flow shortly after the wave crest, which extends over the whole wavelength. Figure 6.1 schematically illustrates characteristic regions of the flow field in the vicinity of the wavy surface reported by Cherukat et al. (1998), and Henn and Sykes (1999). These characteristic regions are the separation region (I), and the regions of maximum positive (II) and maximum negative (III) Reynolds shear stress $-\varrho \overline{u'v'}$. For smooth walls flow-oriented vortical eddies have been associated with large Reynolds stresses, and with the production of turbulence in the viscous region close to the wall (Brooke and Hanratty (1993)). In earlier studies Günther and Rudolf von Rohr (2003), Kruse et al. (2003, 2006), Kuhn et al. (2007), and Wagner et al. (2007) investigated the structure and dynamics of turbulent motions in a developed turbulent flow over various wavy surfaces and identified flow–oriented large–scale structures which contribute most to the momentum transport.

6 Heat Flux in Mixed Convection

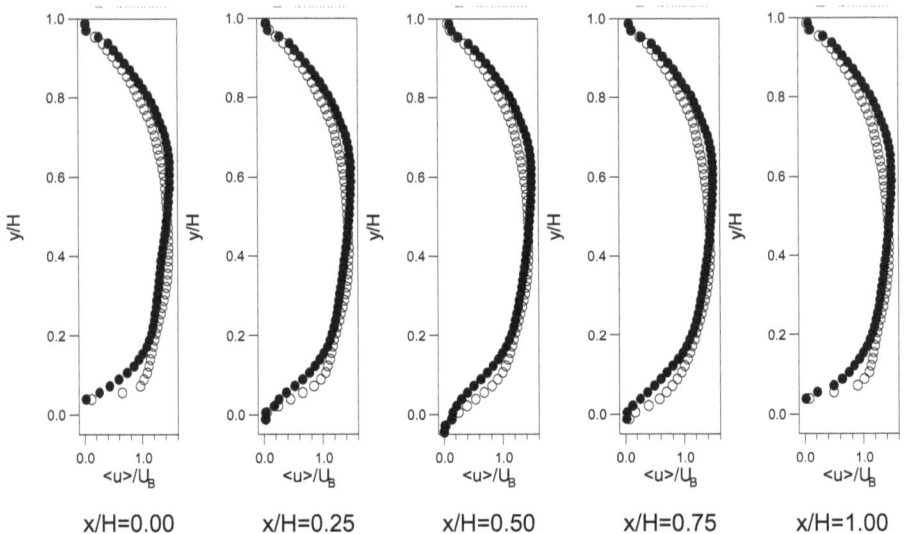

Figure 6.2: Profiles of the mean velocity $\langle u \rangle / U_B$ along one wavelength for Reynolds numbers of $Re_H = 1025$ (unheated •) and $Re_H = 1100$ (heated ○)

The results reported here are for a Reynolds number of $Re_H = 1100$ and a Grashof number of $Gr_H = 1.94 \cdot 10^6$ ($Gr_H/Re_H^2 = 1.6$), and for reference purposes for the isothermal case at $Re_H = 1025$.

6.3 Results

The field of view (FOV) of the measurements in the (x,y)–plane covers the whole region between the wavy surface and the flat top wall (FOV $1.45H$ (streamwise) × $1.12H$ (vertical)). The spatial resolution of the PIV data in this plane of measurement is $0.016H$, which corresponds to 480 µm. The spatial resolution of the PLIF data is $0.009H$, respectively 274 µm.

The field of view for the measurements in the (x,z)–plane is $0.87H$ (streamwise) × $1.14H$ (spanwise), which corresponds to a spatial resolution of the PIV data of $0.013H$, respectively 384 µm. The spatial resolution of the PLIF data in this plane of measurement is $0.0067H$, or 200 µm. The laser light sheet is adjusted to a plane at $y/H = 0.07$ above the wavy surface.

To characterize the influence of the heated bottom surface on the velocity field we plot profiles of turbulence quantities at constant streamwise positions along one wavelength, where $x/H = 0.00$ and $x/H = 1.00$ denotes the wave crest, $x/H = 0.50$ the wave trough.

6.3.1 Mean Velocity Profiles

Figure 6.2 depicts the profiles of the normalized mean velocity $\langle u \rangle / U_B$ for a Reynolds number of $Re_H = 1025$ (unheated •) and $Re_H = 1100$ (heated ○). In the unheated case an asymmetric

6.3 Results

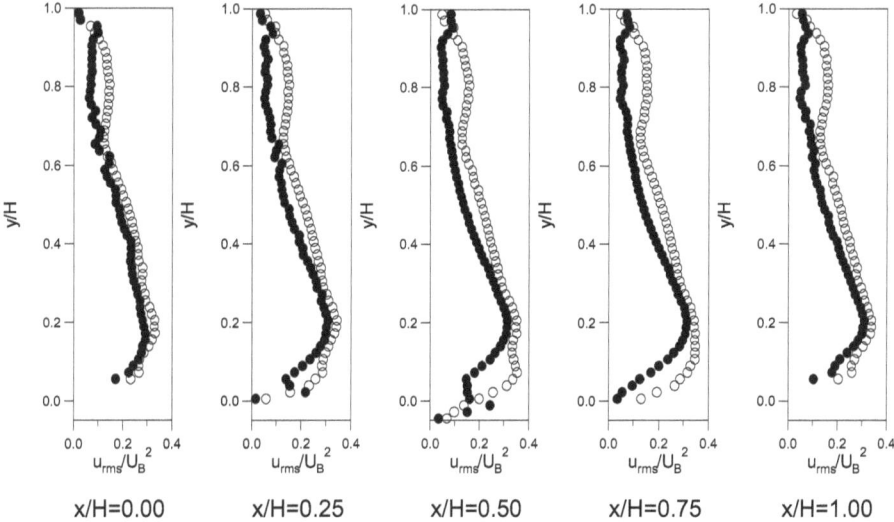

Figure 6.3: Profiles of the root mean square of the streamwise velocity fluctuation $\sqrt{\langle u'^2 \rangle}/U_B$ along one wavelength for Reynolds numbers of $\mathrm{Re}_H = 1025$ (unheated •) and $\mathrm{Re}_H = 1100$ (heated ○)

mean velocity profile is observed. The location of maximum flow velocity is shifted towards the flat top wall and is found at a vertical coordinate of $y/H = 0.65$. This is consistent with earlier studies of isothermal flows over wavy walls at higher Reynolds numbers (e.g. Kruse et al. (2006)). In the heated case also an asymmetric velocity profile is observed, however the location of maximum flow velocity is shifted towards the heated bottom surface ($y/H = 0.4$). In the flow region near the heated surface ($y/H < 0.2$) a higher mean streamwise velocity compared to the unheated case is observed. Negative values of the mean velocity in the region of the wave trough ($x/H = 0.50$) in the heated case indicate a separated flow region, which is not observed in the isothermal case.

6.3.2 Root Mean Square of Velocity Fluctuations

To further characterize the influence of the heated wavy surface we calculate the normalized root mean squares of the streamwise velocity fluctuations $\sqrt{\langle u'^2 \rangle}/U_B$, respectively $\sqrt{\langle v'^2 \rangle}/U_B$ in vertical direction. Figure 6.3 depicts the profiles of the root mean square of the streamwise velocity fluctuation along one wavelength for the Reynolds number $\mathrm{Re}_H = 1025$ (unheated •) and $\mathrm{Re}_H = 1100$ (heated ○). For the heated case in general larger values of the root mean square are found, especially in the region near the heated wavy surface. At the location of the wave crest, i.e. $x/H = 0.00$ and $x/H = 1.00$, the profiles nearly overlap. Downstream and upstream of the wave crest and in the wave trough a maximum is found at a vertical position of approximately $y/H = 0.1$. This indicates that the fluid inside the wave trough accumulates

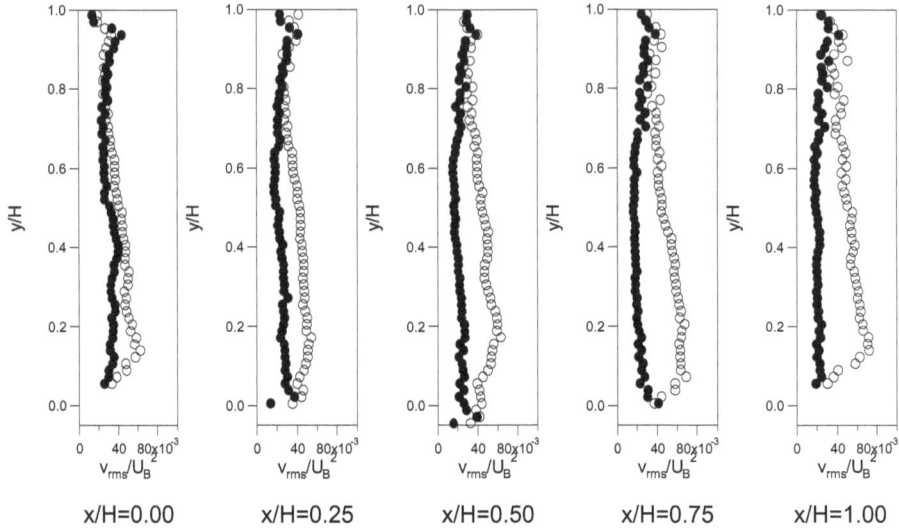

Figure 6.4: Profiles of the root mean square of the vertical velocity fluctuation $\sqrt{\langle v'^2 \rangle}/U_B$ along one wavelength for Reynolds numbers of $\mathrm{Re}_H = 1025$ (unheated •) and $\mathrm{Re}_H = 1100$ (heated ○)

more heat resulting in higher local fluid velocities compared to the isothermal case. A second maximum of the root mean square of the streamwise velocity fluctuations is located in the upper half of the channel near the flat top wall at $y/H \approx 0.8$. This could be an indication of large–scale thermal structures reported in an earlier study (Kruse and Rudolf von Rohr (2006)). Figure 6.4 depicts the profiles of the root mean square of the vertical velocity fluctuations along one wavelength for the Reynolds number $\mathrm{Re}_H = 1025$ (unheated •) and $\mathrm{Re}_H = 1100$ (heated ○). In the profiles the influence of buoyancy due to the resulting fluid motion in vertical direction is observed. In the region between the heated bottom surface and the dimensionless coordinate of $y/H = 0.8$ larger rms values of the vertical velocity fluctuations are found. These results allow to determine the region of the flow which is directly affected by mixed convection.

6.3.3 Reynolds Stress

The flow over waves is associated with a shear layer developing behind the wave crest and extending over the whole wavelength. For mixed convection this shear layer is expected to be intensified through the interaction between the mean flow and the upward fluid motion due to buoyancy. Figure 6.5 depicts the profiles of the Reynolds stress along one wavelength for the Reynolds number $\mathrm{Re}_H = 1025$ (unheated •) and $\mathrm{Re}_H = 1100$ (heated ○). Larger values of the Reynolds stress are found in the region between the heated wavy surface and the location $y/H = 0.6$. The location of the maximum changes from a vertical coordinate of $y/H = 0.15$ for the wave crests to $y/H = 0.10$ in the wave trough. At the upstream side of the wave crest

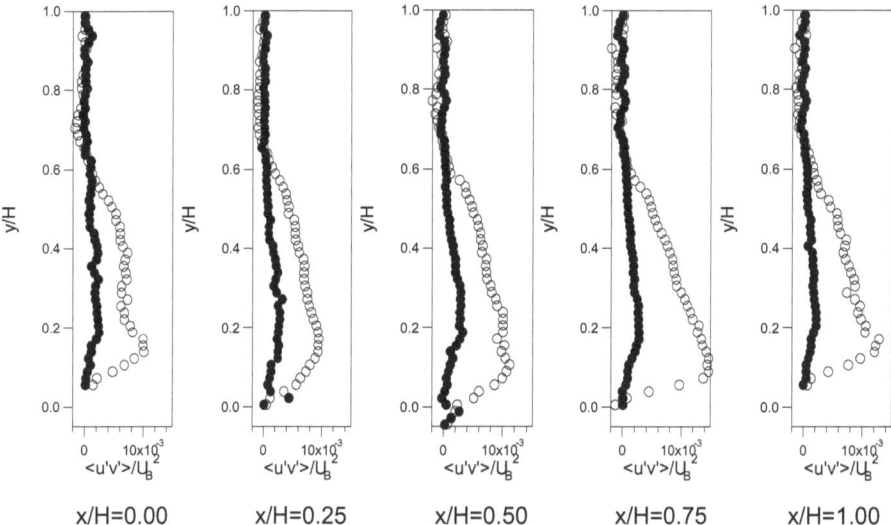

Figure 6.5: Profiles of the Reynolds stress $\langle u'v'\rangle/U_B^2$ along one wavelength for Reynolds numbers of $\mathrm{Re}_H = 1025$ (unheated •) and $\mathrm{Re}_H = 1100$ (heated ○)

($x/H = 0.75$) a pronounced Reynolds stress profile is found. This indicates that the mixed convection is also influenced by the local curvature of the wall.

6.3.4 Mean Temperature Field and Scalar Fluxes

Figure 6.6 depicts the contours of the mean temperature $\langle T \rangle$ normalized with the bulk temperature T_B. The heat transfer from the heated wavy surface to the fluid is observed in the layer of fluid along the surface with elevated temperature. The thickness of the layer increases at the upstream side of the wave, which is consistent with the observations for the velocity field and the turbulence statistics. No temperature changes are observed in the mean temperature field for a region $y/H > 0.2$. To address the transport of temperature we calculate the normalized heat fluxes $\langle u'T'\rangle/(U_B T_B)$, respectively $\langle v'T'\rangle/(U_B T_B)$, from the simultaneous measurements. Figure 6.7 depicts the contours of the magnitude $\left\{ (\langle u'T'\rangle/(U_B T_B))^2 + (\langle v'T'\rangle/(U_B T_B))^2 \right\}^{1/2}$ of the heat flux vector, where dark areas correspond to maxima, respectively white to minima. Near the heated surface the maximum of the flux magnitude is found in a region above the wavy wall ($0.05 < y/H < 0.20$), with a local maximum at the upstream side of the wave. By comparing the flux magnitude with the profiles of the Reynolds shear stress (Figure 6.5) a correlation between the two quantities is observed. The locations of increased momentum transport coincide with the regions of increased scalar transport. The second maximum in the region near the flat top wall $0.85 < y/H < 0.95$ results from transport from the top wall at ambient temperature.

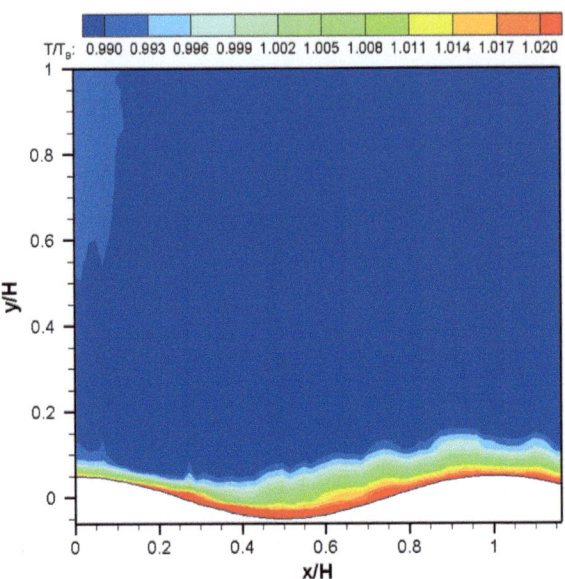

Figure 6.6: Contour of the mean temperature profile $\langle T \rangle / T_B$ in the (x,y)–plane at $\mathrm{Re}_H = 1100$.

6.3.5 Proper Orthogonal Decomposition of the Velocity Field

To extract the most dominant flow structures we perform a proper orthogonal decomposition (POD) of the velocity field according to section 2.4. Figure 6.8 depicts the vector field of the first POD mode for a decomposition of $\mathbf{u}/U_B(x,y/H = 0.10, z, t)$, where $\mathbf{u}$ denotes the measured velocity vector $(u,w)^T$. The contribution of the first POD mode to the total energy of the mixed convective flow is 27.4% which expresses its dominance. Figure 6.9 depicts the fractional and cumulative kinetic energy contribution of the first 15 POD modes. It is observed that the first 5 modes capture nearly 80% of the kinetic energy of the flow. Thus the dominant flow features can be described by addressing only the first few POD modes. The vector field visualizes the structure of the dominant eigenmode as streamwise–oriented, counter–rotating vortices. The spanwise scale, defined as the distance between the cores of the longitudinal flow structures characterized by the same direction of velocity fluctuations, is identified as $1.1H$. This value is smaller than for the first POD mode for the isothermal turbulent flow over waves (Kruse et al. (2006)) and for the heated forced convective flow over waves (Kruse and Rudolf von Rohr (2006)). Thus we identify an influence of mixed convection on longitudinal flow structures present in the flow field.

6.3.6 Scalar Fluxes

Figure 6.10 depicts the contour plot of the magnitude of the heat flux in the (x,z)–plane, where dark areas correspond to maxima, respectively white to minima . By comparing this figure

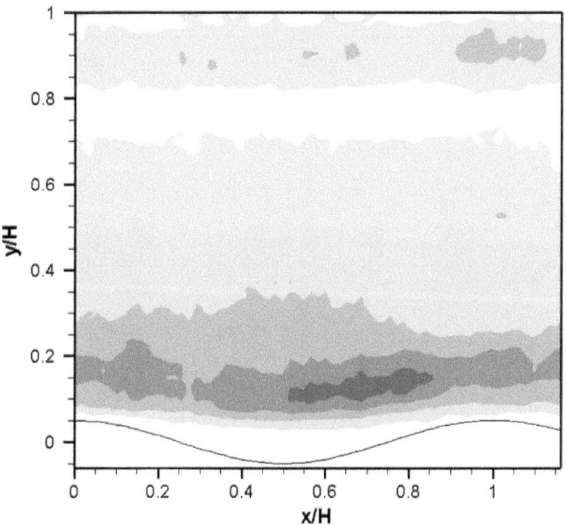

Figure 6.7: Contour plot of the magnitude of the turbulent heat flux in the (x,y)–plane. Dark areas correspond to maximum, light areas to minimum values.

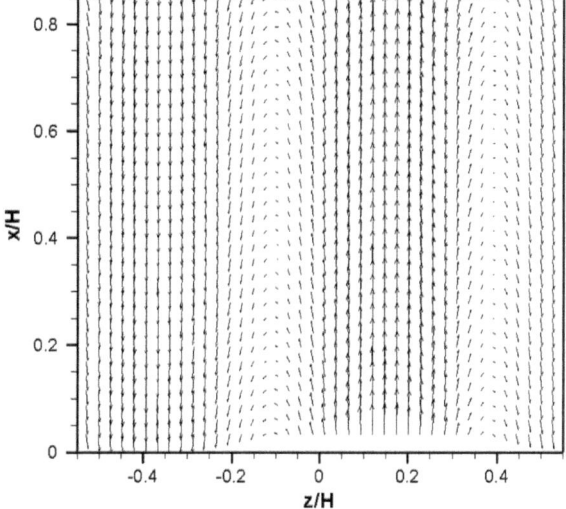

Figure 6.8: Vector field of the first POD mode for a decomposition of $\mathbf{u}/U_B(x,y/H = 0.10,z,t)$ with an energy contribution of 27.4% in the (x,z)–plane (ensemble size 1000 images).

78   6 Heat Flux in Mixed Convection

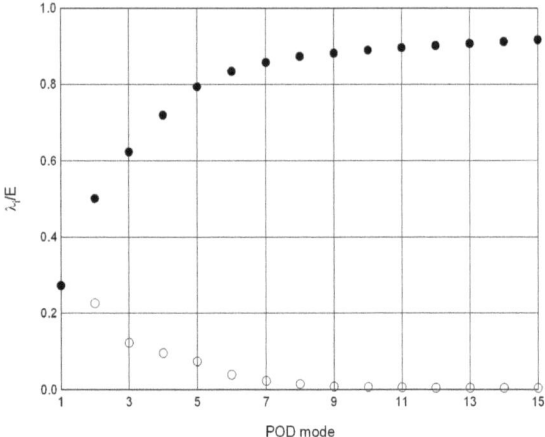

Figure 6.9: Fractional (o) and cumulative (•) kinetic energy contribution from eigenvalues for a decomposition of $\mathbf{u}/U_B(x,y/H = 0.10,z,t)$ (ensemble size 1000 images).

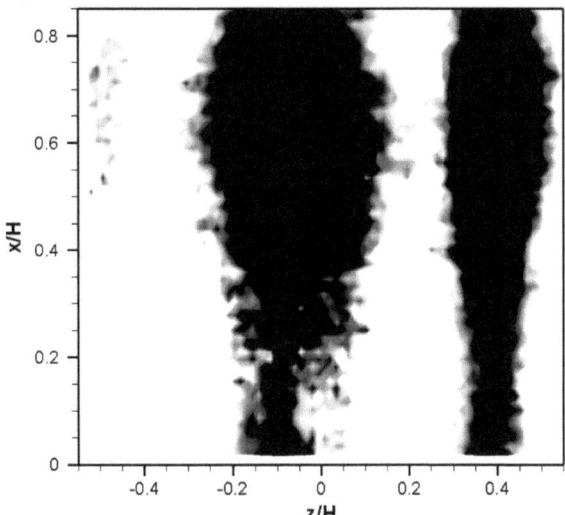

Figure 6.10: Contour plot of the magnitude of the turbulent heat flux in the (x,z)–plane at $y/H = 0.10$. Dark areas correspond to maximum, light areas to minimum values.

6.3 Results

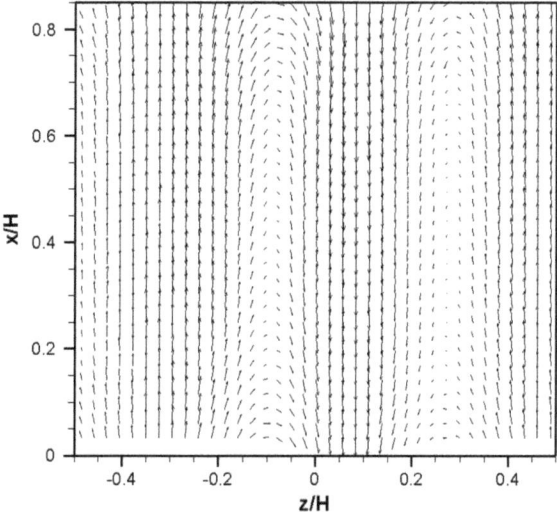

Figure 6.11: Vector field of the first POD mode for a decomposition of $\mathbf{u}T/(U_B T_B)(x,y/H = 0.10, z, t)$ in the (x,z)-plane (ensemble size 1000 images).

with the vector plot of the first POD mode an agreement of the spanwise scales is found. The maximum of scalar transport occurs in the region between the cores of the longitudinal flow structures, i.e. the region of maximum rotation. To further investigate this spatial organization we perform a proper orthogonal decomposition (POD) of the heat flux. Figure 6.11 depicts the vector field of the first POD mode for a decomposition of $\mathbf{u}T/(U_B T_B)(x,y/H = 0.10, z, t)$, where $\mathbf{u}T$ denotes the measured heat flux vector $(uT, wT)^T$. Comparing this result to the first eigenfunction of the decomposition of the velocity field (Figure 6.8) the same spatial organization and the same scaling of $1.1H$ is also observed for the heat flux. Figure 6.12 depicts the fractional and cumulative contribution of the first 15 POD modes. The individual contribution of the eigenvalues for the decomposition of the heat flux is similar to the eigenvalue spectrum of the velocity field (Figure 6.9). The spatial correlation between the momentum field and the scalar field is addressed by applying the proper orthogonal decomposition on the streamwise velocity component and the streamwise heat flux component simultaneously, i.e. to apply the proper orthogonal decomposition on the spatial correlation function between u and uT. Figure 6.13 depicts the dominant eigenfunctions from this decomposition of $((u/U_B),(uT/(U_B T_B)))^T$. Contours of the first two eigenfunctions of the streamwise velocity component are plotted in the first row, the eigenfunctions of the heat flux component are shown in the second row. The spatial organization of the maxima and minima of the eigenfunctions clearly demonstrates the spatial correlation between the streamwise velocity and the streamwise heat flux. Thus the measurements in the (x,z)-plane verify the observations in the (x,y)-plane, locations of increased momentum transport correlate with regions of increased scalar transport. In addition, the scalar transport is increased at the upstream side of the wave $(x/H > 0.50)$, indicated by the broadening of the flux magnitude (Figure 6.10). Thus we also identify an influence of the

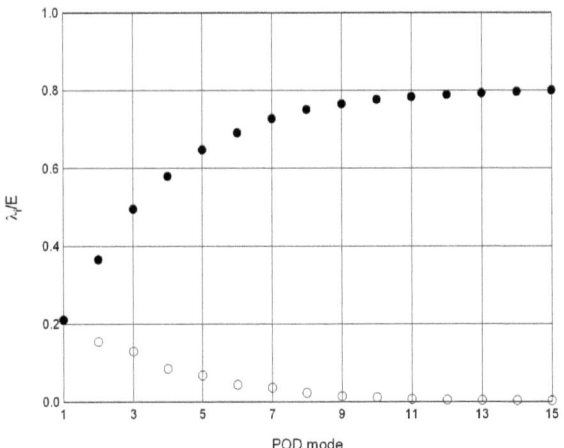

Figure 6.12: Fractional (○) and cumulative (●) contribution from eigenvalues for a decomposition of $\mathbf{u}T/(U_B T_B)$ $(x,y/H = 0.10,z,t)$ (ensemble size 1000 images).

bounding surface geometry on scalar transport processes.

6.4 Conclusions

We apply a combined DPIV/PLIF technique to simultaneously measure the velocity and temperature field of a mixed convective flow over a wavy surface. Therefore we employed two-color PLIF with the fluorescent dye combination Rhodamine B/Rhodamine 110 excited with a Nd:YAG laser. The flow between the flat top and heated wavy bottom surface is investigated at the Reynolds numbers $Re_H = 1100$ (heated), respectively $Re_H = 1025$ (unheated). By recording long image series the statistics of the velocity and temperature field are calculated and we discuss the influence of mixed convection on turbulence quantities and scalar transport properties.

We identified a shift of the mean velocity profile in the vicinity of the heated surface and a separation zone downstream of the wave, both effects differing from the isothermal case. By calculating the root mean square values of the velocity fluctuations and the Reynolds stress we concluded that momentum transport is increased for the mixed convection regime. The maximum of the flux magnitude measured in the (x,y)–plane coincides with the maximum of the Reynolds shear stress. Thus we conclude that locations of increased momentum transport coincide with the regions of increased scalar transport. These observations are verified by the measurements in the (x,z)–plane. There we apply a proper orthogonal decomposition to extract the most dominant flow structures which appear as streamwise–oriented, counter–rotating vortices. Their spanwise scale agrees with the spacing of the maximum heat flux magnitude in this plane of measurement, which is located between the cores of the extracted longitudinal flow structures. In addition we apply the proper orthogonal decomposition on the measured heat flux components. The first eigenfunction of the heat flux and the contribution

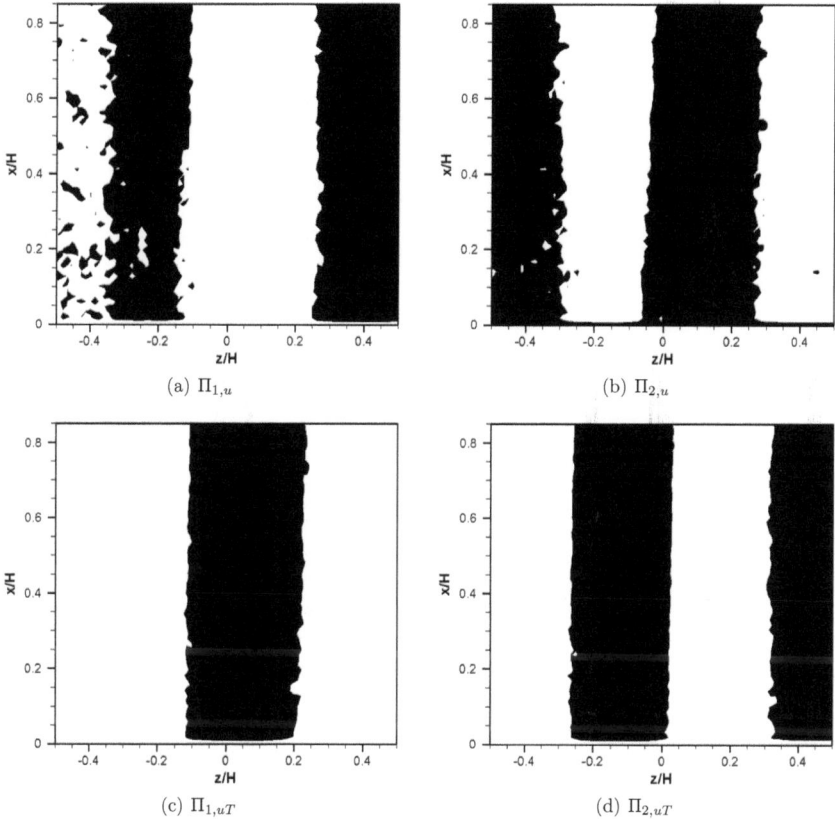

Figure 6.13: Comparison of the first two eigenfunctions for a decomposition of $((u/U_B),(uT/(U_B T_B)))^T (x,y/H = 0.10, z, t)$. The first row depicts the eigenfunctions of the streamwise velocity component, the second row of the streamwise heat flux component (ensemble size 1000 images).

of the individual eigenvalues exhibit similarities to the decomposition of the velocity field. A decomposition of the spatial correlation function between the streamwise velocity component and the streamwise heat flux component revealed the correlation between momentum and heat transport. In addition we find an influence of the bounding surface on the transport processes, visualized by the shape of the mean temperature profile and the increasing flux magnitude in the upstream region of the wave.

By applying a combination of DPIV and two–color PLIF to simultaneously measure the velocity and temperature field we show a correlation of momentum and scalar transport in mixed convective flow over waves.

Chapter 7

The Influence of Mixed Convection on the Transport Properties of a Scalar Species

An experimental study is carried out to address mixed convection from a heated wavy surface. The channel flow between the sinusoidal surface and a flat top wall is investigated by means of a combined digital particle image velocimetry (DPIV) and planar laser–induced fluorescence (PLIF) technique to examine the spatial variation of the streamwise and wall–normal velocity components, and to assess the concentration field of a tracer dye injected into the fluid. We discuss the influence of mixed convection on turbulence quantities and scalar transport properties. Due to the influence of mixed convection we find asymmetric mean velocity profiles and increased momentum transport in the vicinity of the heated surface. The transport of the tracer dye is characterized by enhanced vertical transport due to buoyancy effects and enhanced spanwise transport due to the presence of longitudinal flow structures induced by the convection over the wavy wall. We identify two dominant scalar structures with the most influence on vertical transport. The first one is induced by buoyancy effects, the second one by a combination of buoyancy effects and local wall curvature. Thus the transport properties are additionally enhanced compared to mixed convection from a flat plate by the presence of the wavy surface.

7.1 Introduction

In many transport processes in geophysical and technical flows the transport of a scalar is involved as a combination of free and forced convection. Mixed convective flows are present in technical applications, such as heat transfer devices, and in geophysical flow situations such as transport processes in plant canopies (Banna et al. (2004)). These mixed convective flows are often bound by complex wall geometries, e.g. undulations in heat exchangers to enhance transport processes (e.g. Rush et al. (1999); Dellil et al. (2004)). This experimental study addresses the mixed convective flow between a flat top wall and a heated wavy bottom wall. Simultaneous measurements of the two–dimensional fluid velocity and the scalar field of a tracer emanating from a point source are facilitated from a combined digital particle image velocimetry (DPIV) and high-resolving planar laser–induced fluorescence (PLIF) technique. Thus the effect of mixed convection on transport processes in a complex flow situation, characterized by separation and reattachment of the flow due to the presence of the wavy wall, is evaluated.

Previous studies of mixed convective flows focused mainly on horizontal parallel plate configurations. Osborne and Incropera (1985b) and Osborne and Incropera (1985a) investigated laminar, transitional, and turbulent mixed convection heat transfer for a horizontal parallel plate water channel experimentally. They found an increase in heat transfer due to buoyancy effects which is less pronounced for higher Reynolds numbers. Maughan and Incropera (1989) addressed the regions of heat transfer enhancement for laminar mixed convection for the same flow configuration. Their flow visualization results showed that heat transfer enhancement is preceded by the onset of a secondary flow. Yu, Chang and Lin (1997) and Yu, Chang, Huang and Lin (1997) carried out a numerical and experimental study to investigate this secondary flow in a mixed convective air flow through a horizontal plane channel. Their results showed that this secondary flow is in the form of longitudinal vortex rolls, which changes to transverse rolls when the Reynolds number is lowered (or the Rayleigh number raised). Zhang et al. (2002) addressed the flow patterns and heat transfer enhancement of mixed convective airflow in a rectangular channel. Depending on the Rayleigh number they found two-, four-, and six–roll modes of the longitudinal flow structures, with each mode having an increased effect on heat transfer enhancement. Several other configurations involving planar surfaces have been investigated with regard to scalar transport. Lin and Chen (2006) carried out a numerical study to investigate the effect of rotation on these longitudinal vortices in mixed convective flow over a flat plate. They found that for negative rotation the flow is stabilized since the Coriolis force counteracts the buoyancy force, positive rotation destabilizes the flow. Mixed convection from vertical flat surfaces has also been investigated experimentally (e.g. Ayinde et al. (2006)) and numerically (e.g Angirasa et al. (1997); Evans et al. (2005)), numerical studies also exist for the mixed convection from vertical wavy surfaces (Yao (1983); Moulic and Yao (1989); Jang and Yan (2004)). These studies show lower values for the Nusselt number compared to the corresponding flat plate configuration.

To address the influence of the bounding surface on mixed convective flow in the horizontal configuration wavy walls are chosen as a reference flow case which represent the wall complexity in a well–defined manner. As a test case for the influence of hilly terrain on convection in the atmospheric boundary layer Krettenauer and Schumann (1989) investigated the influence of wavy surfaces on thermal convection by means of direct numerical simulation. Several numerical studies addressed convective heat transfer in channel bound by one or two wavy walls (e.g. Dellil et al. (2004); Metwally and Manglik (2004)). In an experimental study Kruse and Rudolf von Rohr (2006) investigated the transport of heat in a turbulent flow over a heated wavy wall. By employing a particle image thermometry technique the velocity and temperature field was measured simultaneously. Quantitative agreement between large–scale thermal and momentum structures was found. However, all of these studies involving horizontal wavy surfaces were in the forced convection regime, and fluid motion due to buoyancy effects was neglected. The present work addresses these buoyancy effects by applying a combined digital particle image velocimetry and laser–induced fluorescence technique to simultaneously measure the velocity field and the concentration field of a scalar in a channel flow with a heated sinusoidal bottom surface, and thus investigates transport processes in complex mixed convective flows.

Table 7.1 gives an overview over the experiments discussed in this chapter. For reference purposes we investigated two isothermal flow conditions. The two non–isothermal flow conditions in the range of mixed convection are chosen in a way such that for one natural convection is more dominant, while for the other forced convection plays a larger role. The bulk velocities resemble laminar and turbulent flow regimes.

Table 7.1: Investigated flow conditions

Bulk velocity U_B [m/s]	Reynolds number Re_H	Grashof number Gr_H	Ratio $\mathrm{Gr}_H/\mathrm{Re}_H^2$
0.031	1025	0	0
0.033	1100	$1.94 \cdot 10^6$	1.6
0.064	2120	0	0
0.064	2120	$1.30 \cdot 10^6$	0.28

7.2 Results

7.2.1 Velocity Field

The field of view (FOV) of the measurements in the (x,y)–plane covers the whole region between the complex surface and the flat top wall (FOV $1.45H$ (streamwise) $\times$ $1.12H$ (vertical)). The spatial resolution of the PIV data in this plane of measurement is $0.016H$, which corresponds to 480 μm. To characterize the influence of the heated bottom surface on the velocity field we plot profiles of turbulence quantities at constant streamwise positions along one wavelength, where $x/H = 0.00$ and $x/H = 1.00$ denotes the wave crest, $x/H = 0.50$ the wave trough.

Mean Velocity Profiles

Figure 7.1 depicts the profiles of the normalized mean velocity $\langle u \rangle/U_B$ for a Reynolds number of $\mathrm{Re}_H = 1025$ (unheated •) and $\mathrm{Re}_H = 1100$ (heated ○). In the unheated case the mean velocity profile is asymmetric. The location of maximum flow velocity is shifted towards the flat top wall and is found at a vertical coordinate of $y/H = 0.65$. This is consistent with earlier studies of isothermal flows over wavy walls at higher Reynolds numbers (e.g. Kruse et al. (2006)). In the heated case also an asymmetric velocity profile is observed, however the location of maximum flow velocity is shifted towards the heated bottom surface ($y/H = 0.4$). In the complete region of the flow near the heated surface ($y/H < 0.2$) a higher mean streamwise velocity compared to the unheated case is observed. Negative values of the mean velocity in the region of the wave trough ($x/H = 0.50$) in the heated case indicate a separated flow region. This is not found in the isothermal case. Figure 7.2 compares the mean velocity profiles for the unheated (•) and heated (○) case for the Reynolds number $\mathrm{Re}_H = 2120$. For this flow condition the ratio $\mathrm{Gr}_H/\mathrm{Re}_H^2$ equals 0.28, which means that the influence of forced convection is increased. This becomes apparent in the mean velocity profiles which overlap for both cases. The location of maximum flow velocity is found at a vertical coordinate of $y/H = 0.6$, negative mean velocity values at the streamwise position $x/H = 0.50$ indicate flow separation. The only differences in the mean valued are found in the region close to the wall ($y/H < 0.1$). In this near wall region the mean velocity is slightly increased compared to the isothermal case.

Root Mean Square of Velocity Fluctuations

To further characterize the influence of the heated wavy surface we calculate the normalized root mean squares of the velocity fluctuations $\sqrt{\langle u'^2 \rangle}/U_B$, respectively $\sqrt{\langle v'^2 \rangle}/U_B$. Figure 7.3 depicts the profiles of the root mean square of the streamwise velocity fluctuation along one wavelength for the Reynolds number $\mathrm{Re}_H = 1025$ (unheated •) and $\mathrm{Re}_H = 1100$ (heated ○).

86 7 Transport of Species in Mixed Convection

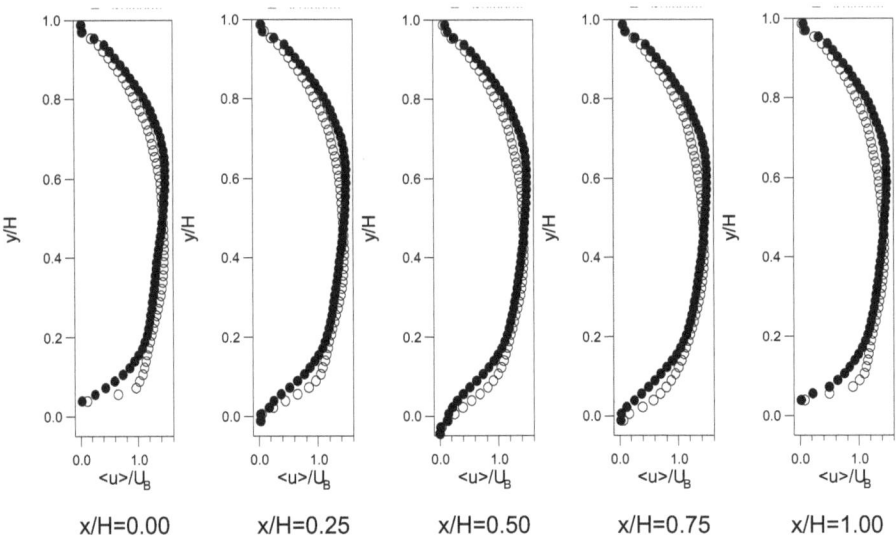

Figure 7.1: Profiles of the mean velocity $\langle u \rangle / U_B$ along one wavelength for Reynolds numbers of $\mathrm{Re}_H = 1025$ (unheated •) and $\mathrm{Re}_H = 1100$ (heated ○)

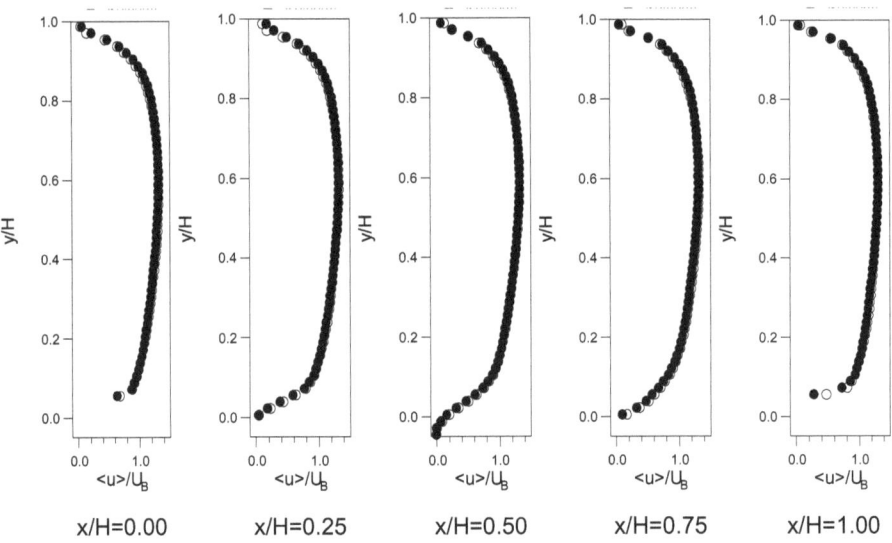

Figure 7.2: Profiles of the mean velocity $\langle u \rangle / U_B$ along one wavelength for a Reynolds number of $\mathrm{Re}_H = 2120$ (unheated •, heated ○)

7.2 Results

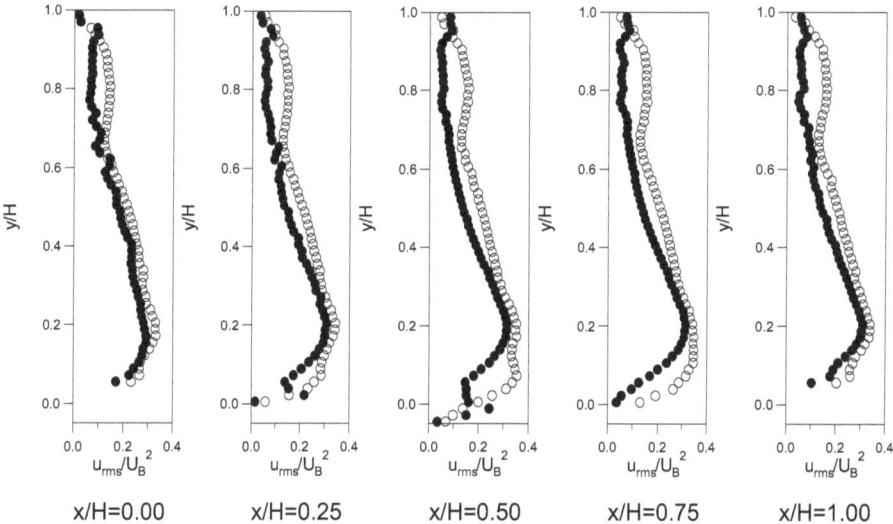

Figure 7.3: Profiles of the root mean square of the streamwise velocity fluctuation $\sqrt{\langle u'^2 \rangle}/U_B$ along one wavelength for Reynolds numbers of $Re_H = 1025$ (unheated •) and $Re_H = 1100$ (heated ∘)

For the heated case in general larger values of the root mean square are found, especially in the region near the heated wavy surface. At the location of the wave crest, i.e. $x/H = 0.00$ and $x/H = 1.00$, the profiles nearly overlap. Downstream and upstream of the wave crest and in the wave trough a maximum is found at a vertical position of approximately $y/H = 0.1$. This indicates that the fluid inside the wave trough accumulates more heat resulting in higher local fluid velocities compared to the isothermal case. A second maximum of the root mean square of the streamwise velocity fluctuations is located in the upper half of the channel near the flat top wall at $y/H \approx 0.8$. This could be an indication of large–scale thermal structures reported in an earlier study (Kruse and Rudolf von Rohr (2006)). Figure 7.4 depicts the profiles of the root mean square of the vertical velocity fluctuations along one wavelength for the Reynolds number $Re_H = 1025$ (unheated •) and $Re_H = 1100$ (heated ∘). In the profiles the influence of buoyancy due to the resulting fluid motion in the vertical direction is observed. In the region between the heated bottom surface and the dimensionless coordinate of $y/H = 0.8$ larger rms values of the vertical velocity fluctuations are found. These results allow to determine the region of the flow which is directly affected by mixed convection. Figs. 7.5 and 7.6 depict the profiles of the root mean square values of the velocity fluctuations for the unheated (•) and heated (∘) case at the Reynolds number $Re_H = 2120$. The profiles nearly overlap, only small deviations due to the influence of mixed convection compared to the turbulent motion induced by the presence of the wavy surface are observed.

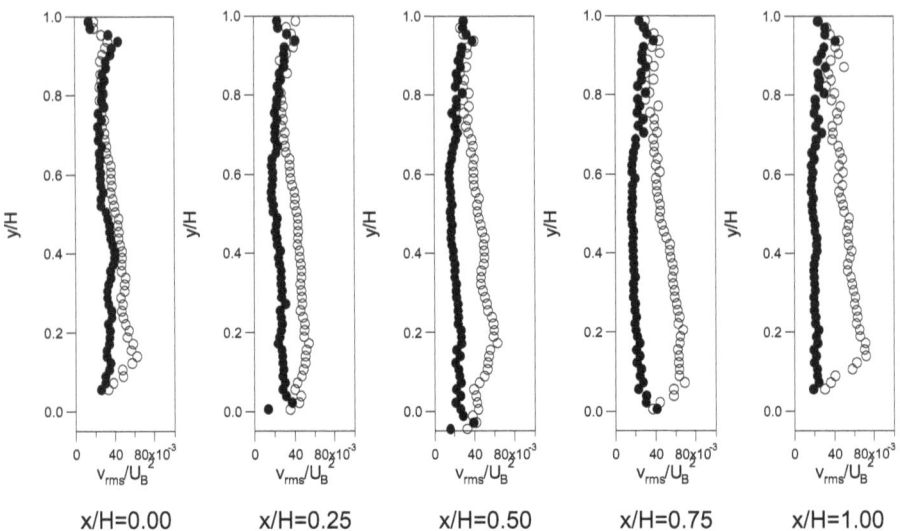

Figure 7.4: Profiles of the root mean square of the vertical velocity fluctuation $\sqrt{\langle v'^2 \rangle}/U_B$ along one wavelength for Reynolds numbers of $\mathrm{Re}_H = 1025$ (unheated •) and $\mathrm{Re}_H = 1100$ (heated ○)

Reynolds Stress

The flow over waves is associated with a shear layer developing after the wave crest and extending over the whole wavelength. For mixed convection this shear layer is expected to be intensified through the interaction between the mean flow and the upward fluid motion due to buoyancy. Figure 7.7 depicts the profiles of the Reynolds stress along one wavelength for the Reynolds number $\mathrm{Re}_H = 1025$ (unheated •) and $\mathrm{Re}_H = 1100$ (heated ○). Larger values of the Reynolds stress are found in the region between the heated wavy surface and the location $y/H = 0.6$. The location of the maximum changes from a vertical coordinate of $y/H = 0.15$ for the wave crests to $y/H = 0.10$ in the wave trough. At the upstream side of the wave crest ($x/H = 0.75$) a pronounced Reynolds stress profile is found. This indicates that the mixed convection is also influenced by the local curvature of the wall. Figure 7.8 compares the Reynolds stress profiles for the unheated (•) and heated (○) case for the Reynolds number $\mathrm{Re}_H = 2120$. For this flow situation the only differences in the Reynolds stresses are found in the region close to the wall ($y/H < 0.2$), where the Reynolds stresses are enlarged.

7.2.2 Scalar Field

The LIF measurements in the (x,y)-plane are performed in the same field of field as the PIV measurements described in the last section (FOV $1.45H$ (streamwise) × $1.12H$ (vertical)). The spatial resolution of the LIF data is $0.008H$, which corresponds to 252 μm. The field of view for the measurements in the (x,z)-plane is $0.87H$ (streamwise) × $1.14H$ (spanwise), which

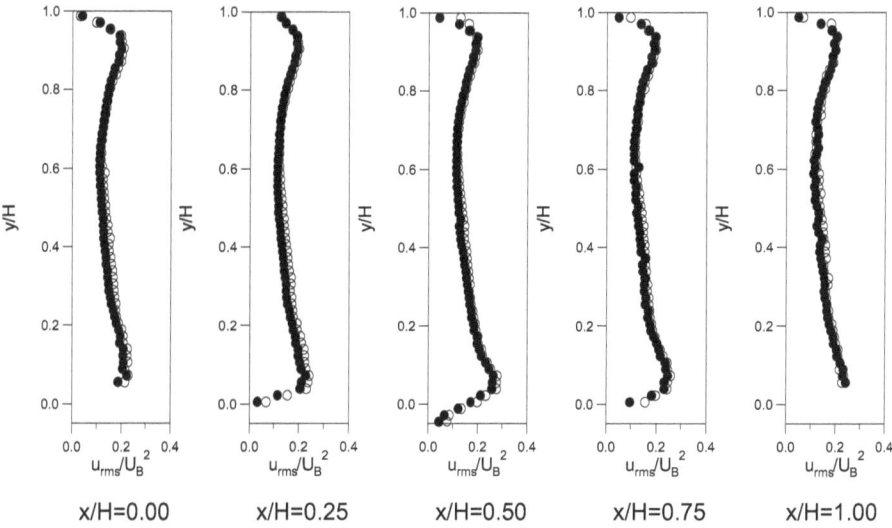

Figure 7.5: Profiles of the root mean square of the streamwise velocity fluctuation $\sqrt{\langle u'^2 \rangle}/U_B$ along one wavelength for a Reynolds number of $Re_H = 2120$ (unheated •, heated ○)

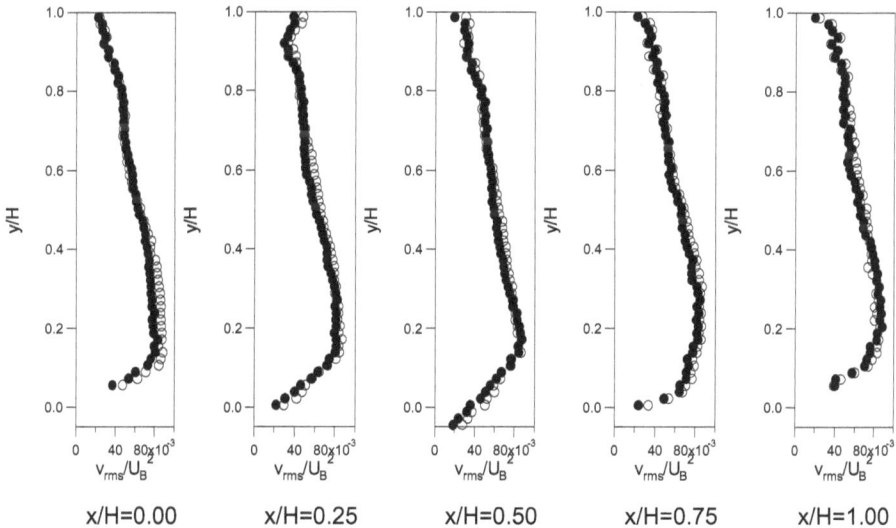

Figure 7.6: Profiles of the root mean square of the vertical velocity fluctuation $\sqrt{\langle v'^2 \rangle}/U_B$ along one wavelength for a Reynolds number of $Re_H = 2120$ (unheated •, heated ○)

90   7 Transport of Species in Mixed Convection

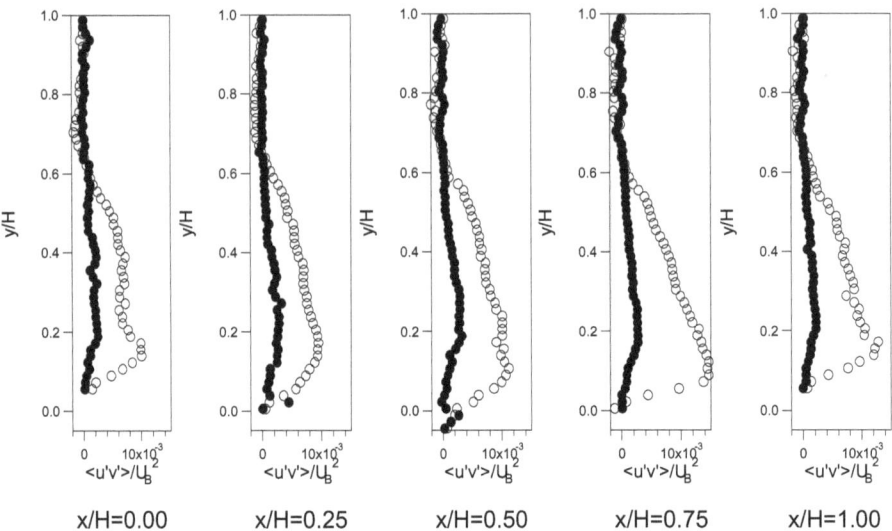

Figure 7.7: Profiles of the Reynolds stress $\langle u'v'\rangle/U_B^2$ along one wavelength for Reynolds numbers of $\mathrm{Re}_H = 1025$ (unheated •) and $\mathrm{Re}_H = 1100$ (heated ∘)

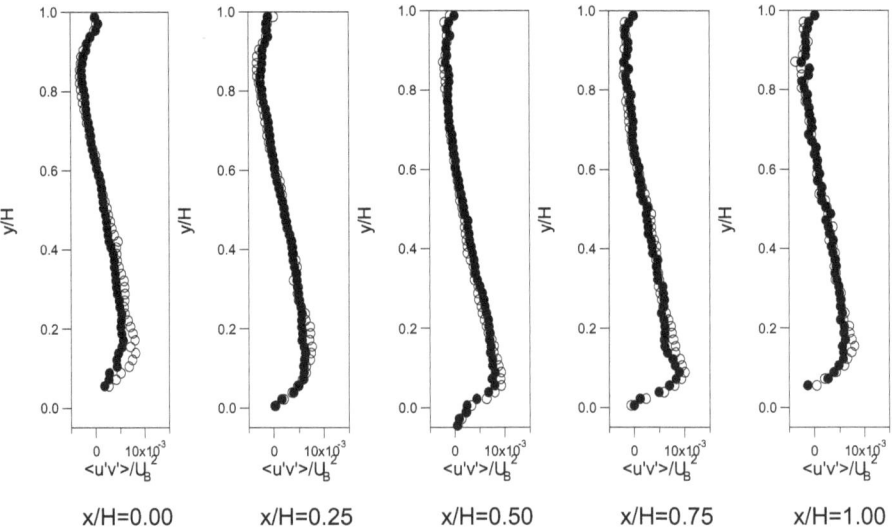

Figure 7.8: Profiles of the Reynolds stress $\langle u'v'\rangle/U_B^2$ along one wavelength for a Reynolds number of $\mathrm{Re}_H = 2120$ (unheated •, heated ∘)

corresponds to a spatial resolution of the LIF data of $0.0067H$, respectively 200 μm. For all measurements the point source is located at the wave crest, all values presented are made dimensionless with the source concentration c_0.

Mean Concentration and Concentration Fluctuation Field in the (x,y)–plane

Figure 7.9 depicts contours of the mean concentration and the root mean square of the concentration fluctuations at Reynolds number $Re_H = 1025$ (unheated, first row), respectively $Re_H = 1100$ (heated, second row). Both, in the mean and the fluctuating concentration field the influence of mixed convection is visible. In the unheated case the scalar is convected with the mean flow, and no transport in vertical direction is observed. This is confirmed by the root mean square values of the concentration fluctuations where also no indication of vertical transport is found. The rather small rms values in the wave trough could result from instantaneous flow separation and the instantaneous formation of a recirculation zone, both not pronounced in the mean flow field. For the heated case the dilution of the mean scalar field along one wavelength is observed. This is an indication of longitudinal flow structures induced by the mixed convection from the heated surface which increase the spanwise transport. The contours of the mean concentration field depict also some amount of the tracer dye being transported into the wave trough. This is an indication of flow separation and the presence of a recirculation zone in the mean flow field. The root mean square values exhibit increased vertical transport at the upstream side of the wave. We will characterize the dominant scalar structures of this vertical transport in the next section. Figure 7.10 depicts contours of the mean concentration and the root mean square of the concentration fluctuations at a Reynolds number of $Re_H = 2120$ for the unheated case (first row) and the heated case (second row). For the unheated case the transport of the scalar with the mean flow is observed, the root mean square values show some transport in vertical direction which is due to the presence of the wavy surface. In the heated case the dilution of the mean concentration field along one wavelength is found as well as the vertical transport of the tracer dye. Compared to the mixed convective flow at a Reynolds number of $Re_H = 1100$ the transport in vertical direction is decreased but still more pronounced than in both unheated cases.

To identify the dominant scalar structures with the largest impact on the vertical transport we investigate instantaneous realizations of the scalar concentration field. Figure 7.11 depicts such instantaneous realizations for $Re_H = 1100$ and $Gr_H = 1.94 \cdot 10^6$ (Figs. 7.11(a) and (b)), and as a reference for $Re_H = 1025$ and $Gr_H = 0$ (Figure 7.11(c)). Figure 7.11(a) depicts a typical ejection of the tracer dye into the outer flow which is also found for mixed convective flows over flat surfaces. Due to buoyancy differences the scalar moves along with packets of rising fluid and is thus transported in vertical direction. Figure 7.11(b) depicts the second prevailing scalar structure. The tracer dye is first trapped in the wave trough and then ejected into the outer flow at the upstream side of the wave. Thus this vertical transport is based on the combination of buoyancy effects and the effects induced by the local wall curvature. An analysis of the image ensemble yields a dominance of the latter described mechanism. Thus we conclude an additional enhancement of transport properties due to the presence of the wavy surface.

Mean Concentration and Concentration Fluctuation Field in the (x,z)–plane

To address the effects of longitudinal flow structures present in the flow field we investigate the spanwise spreading of the scalar plume in the (x,z)–plane. The laser light sheet is adjusted to

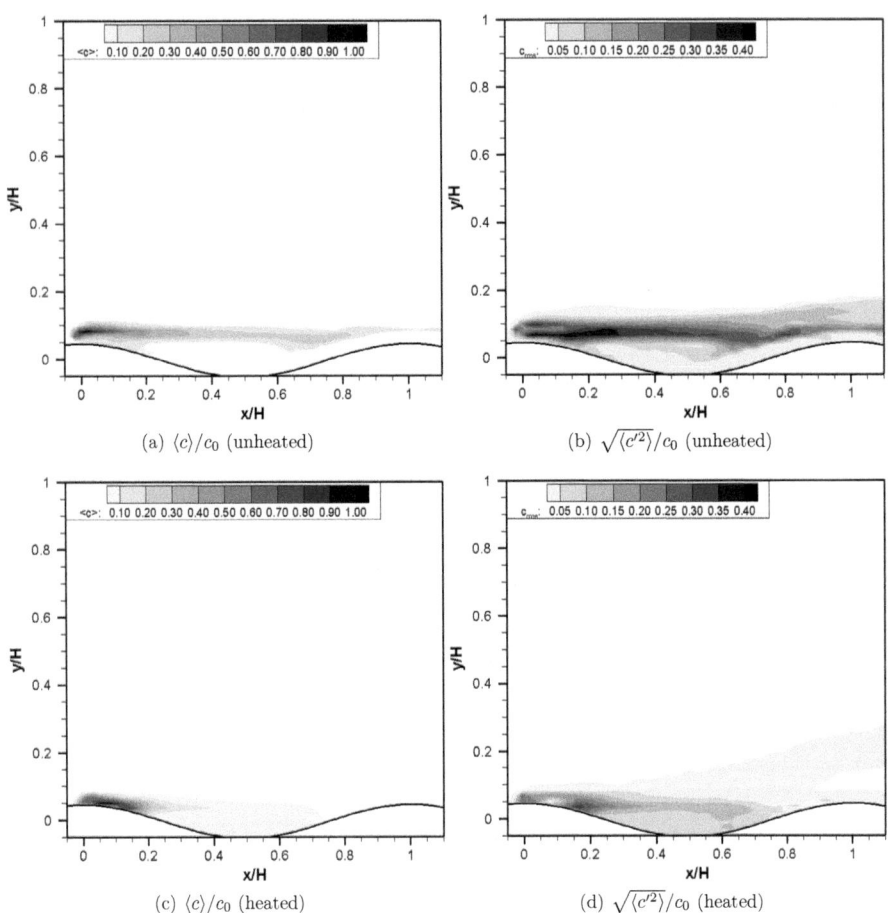

Figure 7.9: Contours of the mean concentration field and the root mean square of the concentration fluctuations for a Reynolds number of $\mathrm{Re}_H = 1025$ (unheated, first row) and of $\mathrm{Re}_H = 1100$ (heated, second row). The point source is located at $x/H = 0.00$, $y/H = 0.05$.

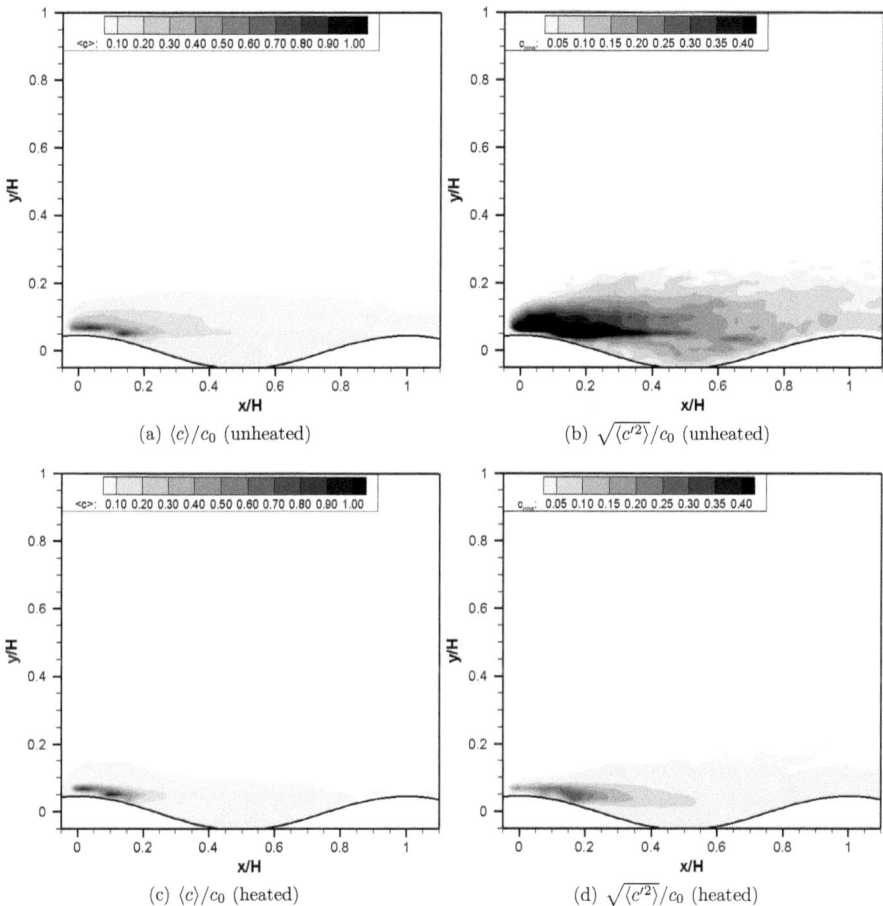

Figure 7.10: Contours of the mean concentration field and the root mean square of the concentration fluctuations for a Reynolds number of $Re_H = 2120$ for the unheated case (first row) and the heated case (second row). The point source is located at $x/H = 0.00$, $y/H = 0.05$.

94   7 Transport of Species in Mixed Convection

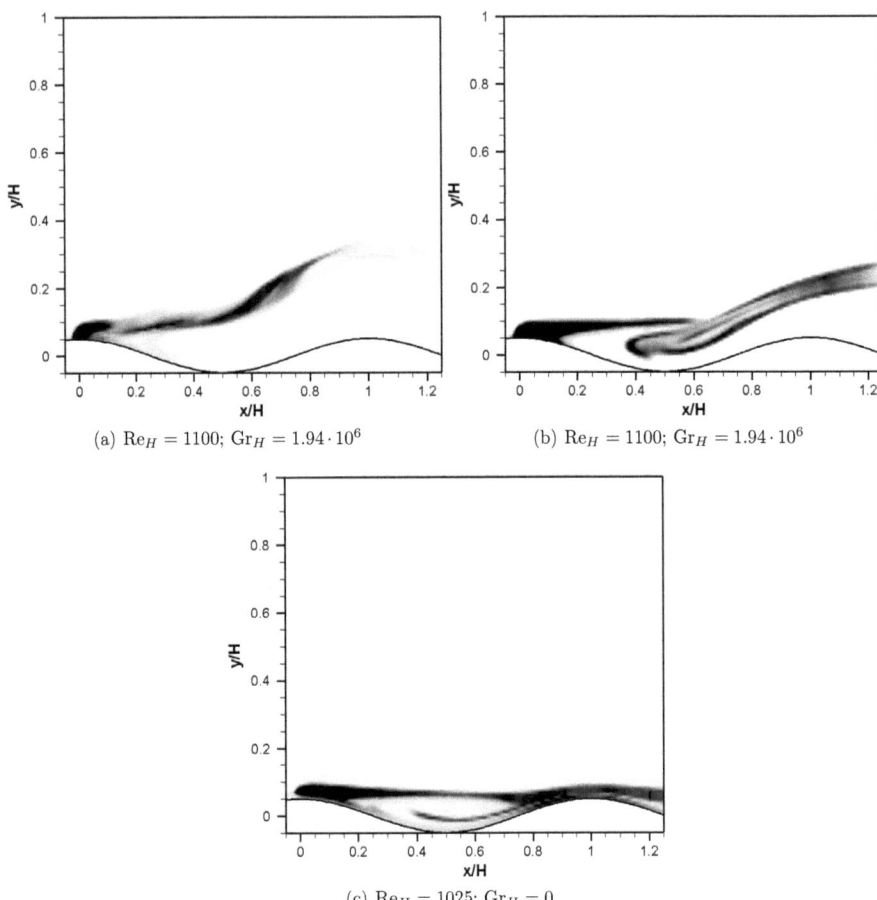

(a) $Re_H = 1100$; $Gr_H = 1.94 \cdot 10^6$

(b) $Re_H = 1100$; $Gr_H = 1.94 \cdot 10^6$

(c) $Re_H = 1025$; $Gr_H = 0$

Figure 7.11: Instantaneous realizations of the scalar concentration field. The point source is located at $x/H = 0.00$, $y/H = 0.05$.

a plane 1 mm above the wavy surface. Figure 7.12 depicts contours of the mean concentration and the root mean square of the concentration fluctuations at Reynolds number $Re_H = 1025$ (unheated, first row), respectively $Re_H = 1100$ (heated, second row). For the unheated case no spanwise transport of the scalar in the mean concentration field is observed. The undulation in the region $0.2 \leq x/H \leq 0.6$ originates from tracer dye which is accumulating in the wave trough and is not due to transport processes. This is consistent with the rms concentration field, where also only a small spanwise spreading is found. In the case of mixed convection a meandering of the scalar plume induced by longitudinal flow structures is observed. This leads to the spanwise spreading of the mean scalar field, and is especially pronounced when calculating the rms concentration field. Thus the spanwise scalar transport is greatly enhanced by mixed convection compared to the isothermal case. Figure 7.13 depicts contours of the mean concentration and the root mean square of the concentration fluctuations at a Reynolds number of $Re_H = 2120$ for the unheated case (first row) and the heated case (second row). For this Reynolds number the flow is in the transition to turbulence, thus also for the unheated case a spanwise transport is observed for both the mean and the rms concentration field. However, this spanwise transport is additionally enhanced by mixed convection. It is worth noting that the rms intensities in both heated cases at $Re_H = 1100$ and $Re_H = 2120$ are comparable. Thus we obtain comparable spanwise transport due to mixed convection which is independent of the flow regime, i.e. if the flow is nominally laminar or turbulent.

7.3 Conclusions

We apply a combined DPIV/PLIF technique to address mixed convection from a wavy surface and the additional transport of a scalar (injected tracer dye). The flow between the flat top and heated wavy bottom surface is investigated at two different Reynolds numbers, in each case heated and isothermal. By recording long image series the statistics of the velocity and the concentration field are calculated and we discuss the influence of mixed convection on turbulence quantities and scalar transport properties.

We identified a shift of the mean velocity profile in the vicinity of the heated surface and a separation zone downstream of the wave, both effects not being present in the isothermal case. By calculating the root mean square values of the velocity fluctuations and the Reynolds stress we concluded that momentum transport is increased for the mixed convection regime. The transport of the tracer dye is characterized by enhanced vertical transport due to buoyancy effects and enhanced spanwise transport due to the presence of longitudinal flow structures induced by the mixed convection. This spanwise transport in mixed convection is comparable for laminar and turbulent flow conditions. By analyzing instantaneous realizations of the scalar concentration field we identify two dominant scalar structures contributing most to vertical transport. The first one is characterized by the vertical transport of the tracer dye with rising packets of fluid with decreased density, i.e. induced by buoyancy effects. The second one results from a combination of buoyancy effects and local wall curvature. The tracer dye is trapped in the wave trough and is then transported into the outer flow at the upstream side of the wave. The latter transport mechanism occurs at a higher frequency compared to the first. Thus the transport properties are additionally enhanced compared to mixed convection from a flat plate by the presence of the wavy surface.

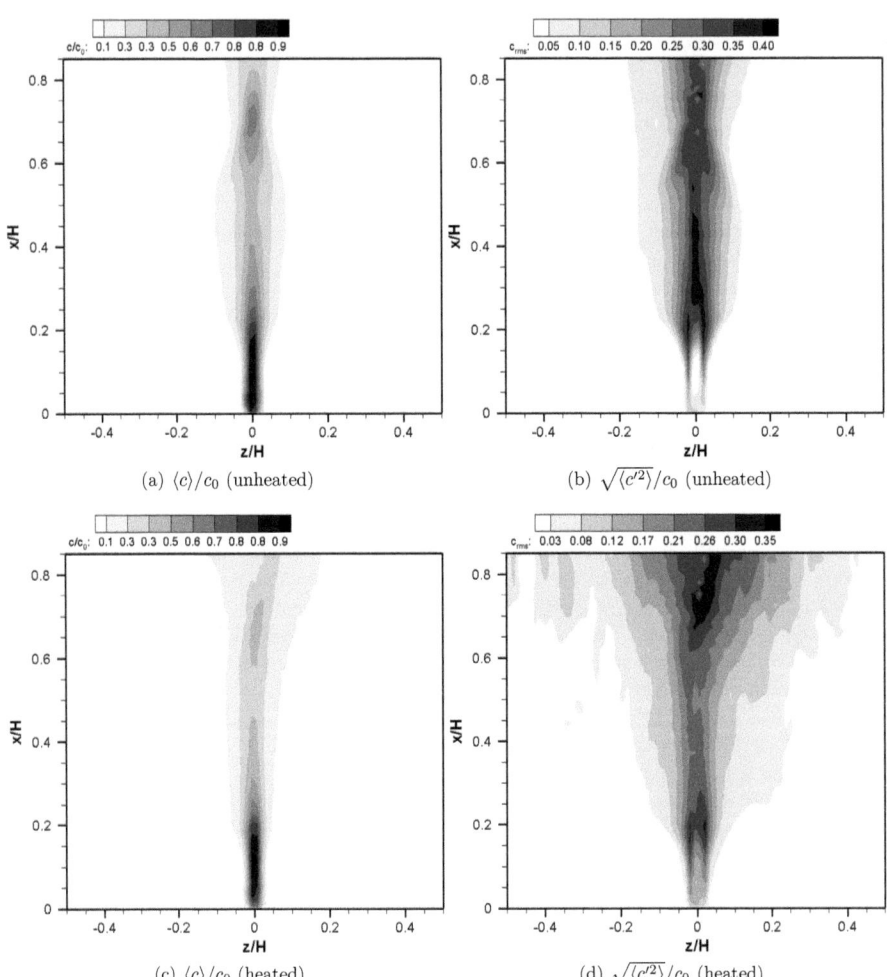

Figure 7.12: Contours of the mean concentration field and the root mean square of the concentration fluctuations for a Reynolds number of $\mathrm{Re}_H = 1025$ (unheated, first row) and of $\mathrm{Re}_H = 1100$ (heated, second row). The point source is located at $x/H = 0.00$, $z/H = 0.00$.

7.3 Conclusions

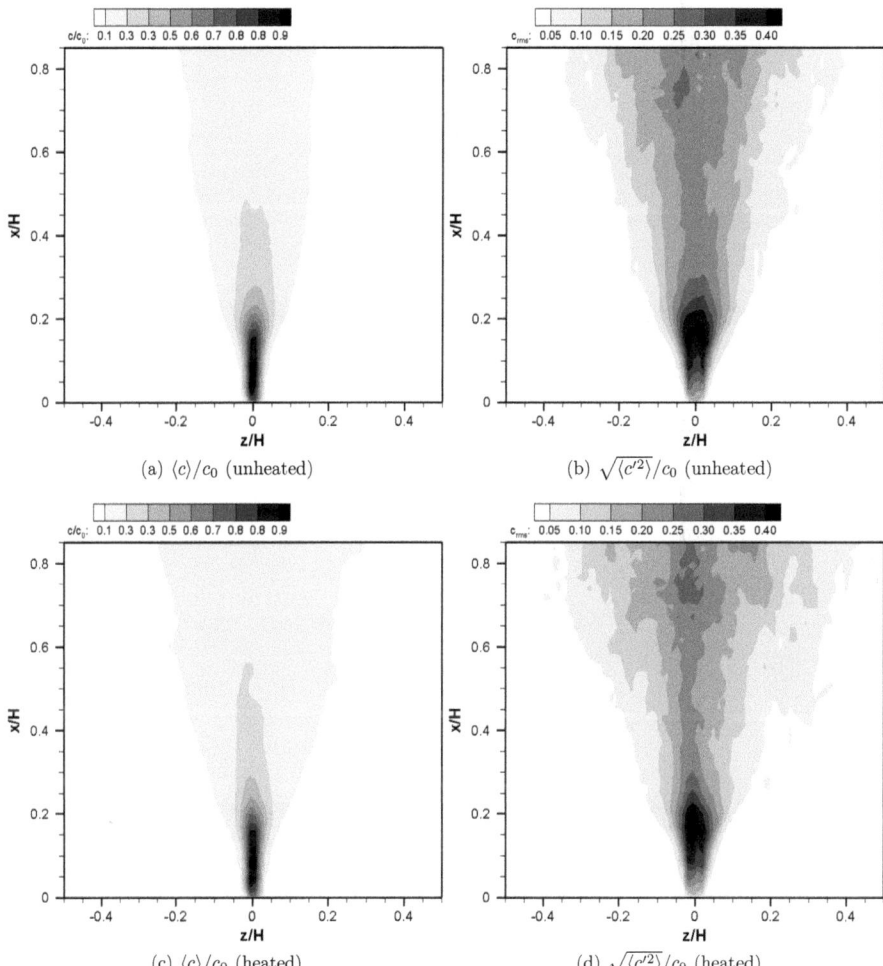

Figure 7.13: Contours of the mean concentration field and the root mean square of the concentration fluctuations for a Reynolds number of $Re_H = 2120$ for the unheated case (first row) and the heated case (second row). The point source is located at $x/H = 0.00$, $z/H = 0.00$.

Numerical Investigation of Mixed Convective Flows

Chapter 8
Numerical Simulation of Mixed Convection over a Wavy Wall: A Dynamical LES Approach

In this numerical study we investigate a mixed convective flow of water over a heated wavy surface over a range of Reynolds numbers from $Re_H = 20$ to 2000 and Richardson numbers from $Ri = 0.5$ to 5000. With this transitional and turbulent flow regimes are addressed. A dynamic Large Eddy Simulation (LES) approach is applied where the thermal buoyancy effects are represented by the Boussinesq approximation. The LES results show good agreement with available measurements including first and second order statistics of velocity and thermal fields. We focus our investigation on the thermal buoyancy effects on the wall heat transfer and the spatial reorganization of the vortical flow structures. In order to characterize the reorganization of the mean flow features, the vortical coherent structures are identified and extracted according to the swirling strength criteria. Interesting reorganization of flow structures takes place between $Re_H = 20$ and $Re_H = 200$ where the initially spanwise oriented large coherent structures start to be streamwise oriented. With further increase of Re_H, these large structures disappear from the central part of the simulated domain and reappear in the proximity of the horizontal wavy wall for $Re_H \geq 1000$. The imprints of this flow reorganization are clearly visible on distributions of the local heat transfer coefficient (Nusselt number) along the horizontal wavy wall. The integral heat transfer for the wavy wall configuration is significantly enhanced (≈ 2.5 times) for $Re_H = 1000, 2000$ in comparison with the standard flat horizontal wall configuration.

8.1 Introduction

Mixed convective flows are present in technical applications, such as heat transfer devices, and in geophysical flow situations such as transport processes in plant canopies, e.g. Banna et al. (2004). These mixed convective flows are often bound by complex wall geometries, e.g. undulations in heat exchangers to enhance transport processes (e.g. Rush et al. (1999); Dellil et al. (2004)). In this chapter we numerically investigate the mixed convective channel flow between a flat top wall and a heated wavy bottom wall at Richardson numbers ranging from $Ri = Gr_H/Re_H^2 = 0.5$ to 5000. We address the effect of buoyancy induced fluid motions together with the influence of the wavy bounding surface on the mean flow and wall heat transfer. We compare our numerical predictions with experimental results of a wavy channel flow facility with the same geometrical configuration and boundary conditions.

Previous experimental studies of mixed convective channel flows focused mainly on horizontal parallel plate configurations. Osborne and Incropera (1985a) and Osborne and Incropera

(1985b) experimentally investigated laminar, transitional, and turbulent mixed convection heat transfer for a horizontal parallel plate water channel. Their studies covered Reynolds numbers in the range of 65≤Re≤6500, and Grashof numbers between $4.2 \cdot 10^5 \leq Gr \leq 2.8 \cdot 10^8$. They found an increase in heat transfer due to buoyancy effects by a factor of 6 in the laminar flow regime. This heat transfer enhancement was significantly reduced for higher values of Reynolds numbers where a factor of 1.3 was obtained. Maughan and Incropera (1989) analyzed the regions of heat transfer enhancement for laminar mixed convection for the same flow configuration with $2 \cdot 10^4 \leq Gr \leq 8 \cdot 10^5$ at constant Reynolds numbers of Re = 125, 250, or 500, respectively. Their flow visualization results showed that heat transfer enhancement for all cases is preceded by the onset of a secondary flow. Yu, Chang and Lin (1997) and Yu, Chang, Huang and Lin (1997), carried out a numerical and experimental study to investigate this secondary flow in a mixed convective air flow through a horizontal plane channel at Reynolds number Re≤50 and 3000≤Ra≤10000. Their results showed that this secondary flow is in the form of longitudinal vortex rolls, which changes to transverse rolls when the Reynolds number is lowered (or the Rayleigh number increased). For Re<50 they observed only transverse structures, i.e. that thermal buoyancy effects dominate the mean flow. Zhang et al. (2002) addressed the flow patterns and heat transfer enhancement of mixed convective airflow in a rectangular channel at Re=40 and 100≤Ra≤4200. Depending on the Rayleigh number they found two– (Ra = 100), four– (Ra ≥ 630), and six–roll (Ra ≥ 1708) modes of the longitudinal flow structures. Each roll mode produced significant heat transfer enhancement.

An interesting practical application was addressed in Alawadhi (2005) where the convective cooling enhancement for rectangular blocks using a wavy plate was investigated. This problem is particularly important for the efficient cooling of the printed circuit board assembly. The two–dimensional numerical simulations were performed for different plate waviness at Reynolds numbers in a range of 250≤Re≤1000. The numerical simulation demonstrated that the wavy plate enhanced cooling of the rectangular blocks and reduced their surface temperature up to 25%.

The forced convective flow over a wavy wall in fully devolved turbulent flow regimes has been extensively studied in literature. In a direct numerical simulation (DNS) Cherukat et al. (1998) addressed the characteristics of the developing shear layer above the recirculation zone formed behind the wave crest of a wavy wall with an amplitude–to–wavelength ratio of $\alpha = 0.05$ at a Reynolds number of 3460 (defined with the bulk velocity and the half channel height). Large eddy simulations (LES) of forced convective wavy boundary flow were performed by Henn and Sykes (1999), and Tseng and Ferziger (2004). In their extensive study Henn and Sykes (1999) focused on the effects of different wave slopes ranging from 0 to 0.628 on turbulence at 5720≤Re≤20060.[1] They identified a scaling between the slope of small–amplitude waves and the velocity fluctuations, which is linear for the streamwise and vertical velocity components, but squared for the lateral. This is caused by the presence of coherent structures located at the upstream side of the wave. Tseng and Ferziger concentrated on these coherent structures for a wavy wall of α=0.1 at Re=2400, and illustrated the vortex formation and transport of Görtler vortices induced by the local wall curvature.

The buoyancy driven convection from wavy surfaces was investigated by Krettenauer and Schumann (1989) by means of a DNS, and by Hanjalić and Kenjereš (2001) by means of T–RANS. Krettenauer and Schumann investigated Rayleigh numbers up to 4.5×10^4, and they observed that the flow features were not sensitive to alterations of the wavy surface. In their

---

[1] The wave slope is given by ak, with the wavenumber $k = 2\pi/\Lambda$.

approach Hanjalić and Kenjereš investigated the same geometrical configuration at increased Rayleigh numbers up to Ra = 10^9 and identified changes in the organization of the coherent structures when advancing from two- to three-dimensional waviness. The forced convective flow over different wavy walls having amplitude-to-wavelength ratios of 0.01, 0.05, and 0.1 at a Reynolds number of Re = 6760 was addressed by Choi and Suzuki (2005) by means of LES. An increase in heat transfer efficiency depending on the geometrical parameters of the wavy surface was observed. By changing the amplitude-to-wavelength ratios from 0.01 to 0.1 the Nusselt numbers increased by a factor of 4. This heat transfer enhancement is the topic of several other numerical works employing turbulence models addressing convective heat transfer in channel bound by one or two wavy walls (e.g. Dellil et al. (2004),Metwally and Manglik (2004)).

It is important to note that the great majority of previous experimental and numerical studies in literature focused on either forced or natural convection situations over wavy walls. The primary objective of this study is to fill this gap and to focus on the mixed convection case for the wavy wall configuration. Additional motivation lies in the possibility to perform a direct validation of the numerical simulations with experimental studies. In the present numerical work we specifically address the influence of thermal buoyancy in the mixed convective flow over a wavy wall at different Richardson numbers (with the Reynolds number ranging from Re_H = 20 to Re_H = 2000). By performing a dynamic LES, we focused our investigations on the wall heat transfer and reorganization of the flow structures.

8.2 Flow Description

We consider the mixed convective flow between a flat top wall and a heated sinusoidal bottom surface. The bottom surface is characterized by the amplitude-to-wavelength ratio of $\alpha = 2a/\Lambda = 0.1$, having a wavelength of $\Lambda = 30$ mm which is equal to the channel height H. The profile of the wavy wall and the coordinate system used is depicted in Figure 8.1. The profile of the wavy wall can be described by

$$y_w(x) = a \cos\left(\frac{2\pi x}{\Lambda}\right). \tag{8.1}$$

The mixed convective flow is characterized by the Reynolds number and the Rayleigh number. The Reynolds number Re_H is calculated according to

$$Re_H = \frac{U_B H}{\nu}, \tag{8.2}$$

where ν denotes the kinematic viscosity, H is the height of the channel, and the bulk velocity U_B is given by

$$U_B = \frac{1}{H - y_w} \int_{y_w}^{H} U(x_\xi, y) \, dy, \tag{8.3}$$

where x_ξ denotes an arbitrary x-location and y_w describes the profile of the complex surface. The Rayleigh number Ra_H is given by

$$Ra_H = \frac{g H^3 \Delta T \beta}{\nu^2} Pr, \tag{8.4}$$

where ΔT denotes the temperature difference between the bottom and top wall, β is the volumetric thermal expansion coefficient, and Pr denotes the Prandtl number of the fluid. The

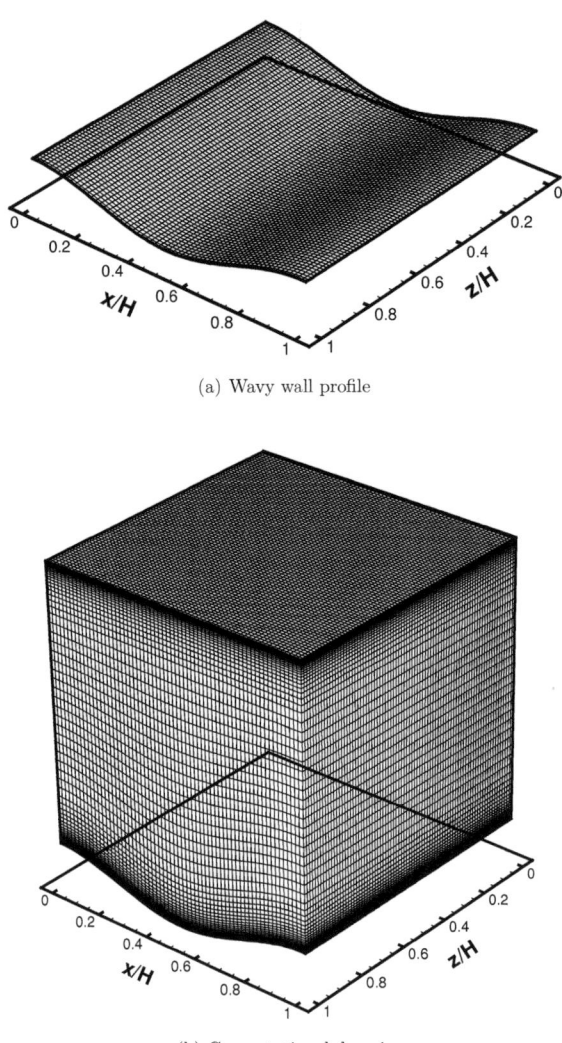

(a) Wavy wall profile

(b) Computational domain

Figure 8.1: Computational domain and profile of the wavy wall characterized by $\alpha = 0.1$ and $\Lambda = 30$ mm

Table 8.1: Large eddy simulations of mixed convective flow over waves.

Re_H	Ra_H	Ri	N_x	N_y	N_z	Δx^+	y^+	Δz^+
20	$1.4 \cdot 10^7$	5000	64	64	80	0.43	0.02	0.34
200	$1.4 \cdot 10^7$	50	64	64	80	1.03	0.06	0.83
1000	$1.4 \cdot 10^7$	2	64	64	80	1.88	0.11	1.50
2000	$1.4 \cdot 10^7$	0.5	64	64	80	3.44	0.20	2.75

details of the simulations reported in this chapter and the specifications of the numerical mesh (number of grid points in each coordinate direction, N_i, and the grid spacing in wall units, Δx_i^+) are tabulated in Table 8.1. For the simulations we apply periodic boundary conditions in the streamwise and spanwise direction by defining the total mass flux in the streamwise direction through the domain. The top wall is kept at a constant temperature ($T = 20$ °C), while on the bottom wall a constant heat flux boundary condition ($\dot{q} = 1800$ W/m^2) is imposed.

8.3 Subgrid Models and Numerical Details

The equations describing turbulent flows and heat transfer are given by the conservation of mass, momentum and thermal energy. The spatially filtered form of the conservation of momentum equation can be written as

$$\frac{\partial (\varrho \langle u_i \rangle)}{\partial t} + \frac{\partial (\varrho \langle u_j \rangle \langle u_i \rangle)}{\partial x_j} = -\frac{\partial \langle p \rangle}{\partial x_i} + \mu \frac{\partial^2 \varrho \langle u_i \rangle}{\partial x_j \partial x_j} - \frac{\partial \varrho \tau_{ij}^R}{\partial x_j} + F_i, \quad (8.5)$$

and the conservation of thermal energy equation becomes

$$\frac{\partial (\varrho \langle T \rangle)}{\partial t} + \frac{\partial (\varrho \langle u_j \rangle \langle T \rangle)}{\partial x_j} = \frac{\partial}{\partial x_j} \left[\frac{\mu}{\Pr} \left(\frac{\partial \langle T \rangle}{\partial x_j} \right) - \varrho \tau_{\theta j} \right], \quad (8.6)$$

where $\bar{p}$ is the filtered pressure field, F_i is an external force, μ denotes the dynamic viscosity of the fluid, Pr is the Prandtl number defined as $\Pr = \frac{\mu c_p}{\lambda}$, and $\tau_{ij}^R, \tau_{\theta j}$ represent the unresolved turbulent stress and turbulent heat flux, respectively. For more details about the filtering procedure we refer to section A.1. The coupling between the temperature and the velocity fields is accomplished by applying the Boussinesq approximation. In this approximation the density changes of the fluid due to a temperature gradient are assumed to be negligible, and the resulting buoyancy motion is introduced as external force F_B in the momentum equation, which reads

$$F_B = -\varrho g_i \beta (T - T_{ref}), \quad (8.7)$$

where β is the thermal expansion coefficient of the fluid, and g_i the acceleration due to gravity.

The large eddy simulations are performed by using the dynamic Smagorinsky model for the subgrid scales as described by Germano et al. (1991). The unresolved traceless subscale stresses τ_{ij}^R are related to the rate of strain $\langle S_{ij} \rangle$ of the resolved velocity field by employing the Boussinesq eddy–viscosity concept

$$\tau_{ij}^R - \frac{1}{3} \tau_{kk} \delta_{ij} = -2 \nu_t \langle S_{ij} \rangle. \quad (8.8)$$

The eddy viscosity is defined as

$$\nu_t = (C_S \Delta)^2 |\langle S \rangle|, \tag{8.9}$$

where Δ denotes the length scale of the unresolved motion calculated from the volume of the computational cell ΔV

$$\Delta = (\Delta V)^{1/3}, \tag{8.10}$$

and $|\langle S \rangle|$ is the magnitude of the strain rate defined as

$$|\langle S \rangle| = \sqrt{2 \langle S_{ij} \rangle \langle S_{ij} \rangle}. \tag{8.11}$$

C_S is evaluated from the dynamical procedure

$$C_S = -\frac{1}{2} \frac{L_{ij} M_{ij}}{M_{kl} M_{kl}}, \tag{8.12}$$

where $L_{ij} = \widehat{\langle u_i \rangle \langle u_j \rangle} - \widehat{\langle u_i \rangle} \, \widehat{\langle u_j \rangle}$ represents the resolved turbulent stress of the scales between Δ and a coarse $\widehat{\Delta}$ (where $\widehat{\Delta} = 2\Delta$), and $M_{ij} = \widehat{\Delta}^2 \widehat{|\langle S \rangle| \langle S_{ij} \rangle} - \Delta^2 \widehat{|\langle S \rangle|} \langle S_{ij} \rangle$ represents the contribution of the modeled stress of those scales.

For the simulations a fully unstructured, finite–volume based second order accurate numerical code is used (Ničeno (2001)). Both, convective and diffusive terms of the discredited equations are approximated by a second-order central difference scheme (CDS). Since the wavy surface geometry is easy to be represented, hexagonal control volumes are employed for the numerical mesh. The employed numerical mesh is refined in the proximity of the thermally active horizontal walls in order to fully resolve both hydrodynamical and thermal boundary layers. Details of the numerical mesh are provided in Table 8.1. The used numerical mesh was fully appropriate for the LES approach since the maximum of the ratio between the subgrid turbulent and the molecular viscosity for the highest value of $Re_H = 2000$ was below 0.5. This can be additionally confirmed by plotting vertical profiles of the ratio between the characteristic mesh size and the Kolmogorov length scale at $x/H = 0.5$, as shown in Figure 8.2. It can be concluded that a well–resolved LES is achieved for all considered values of Re_H. The time integration is performed by a fully implicit second–order three–consecutive time–steps method. The value of the characteristic time step is selected to ensure that at each particular time instant the value of $CFL_{max} < 0.5$. The time averaging started after reaching a fully developed turbulent regime (usually reached after 4000 time steps) and continued for additional 30000 time steps.

The dynamic procedure to obtain the Smagorinsky constant ensures correct near wall behavior of the turbulent viscosity eliminating necessity to include the wall–damping functions. Figure 8.3(a) shows the time–dependent behavior of the Smagorinsky constant C_S at two characteristic monitoring points located in the center of the simulated domain and in the proximity of the bottom wall, respectively. Similarly, the vertical profiles of C_S at different locations along the streamwise direction are shown in Figure 8.3(b).

The subgrid turbulent heat flux is calculated from the simple gradient diffusion hypothesis as

$$\tau_{\theta i} = \frac{\nu_t}{\Pr_t} \frac{\partial \langle T \rangle}{\partial x_i}, \tag{8.13}$$

where $\Pr_t = 0.9$ is the turbulent Prandtl number.

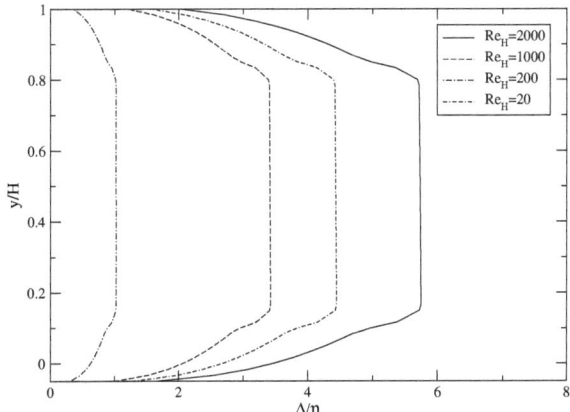

Figure 8.2: Vertical profiles of the ratio between characteristic control volume ($\Delta=\Delta V^{1/3}$) and Kolmogorov ($\eta = (\nu^3/\varepsilon)^{1/4}$, $\varepsilon = \overline{u}_{max}^3/H$) length scales at the location $x/H = 0.5$, for $\text{Re}_H = 20, 200, 1000, 2000$, respectively.

8.4 Flow Field

8.4.1 Velocity Profiles

In this section the influence of the different mixed convective regimes on the mean velocity profile in the (x,y)-plane is discussed. Therefore we plot vertical velocity profiles at distinct streamwise locations x/H along the complex surface, where $x/H = 0.00$ and $x/H = 1.00$ denote the wave crest, $x/H = 0.50$ the wave trough, and $x/H = 0.25$ and $x/H = 0.75$ the inflection point of the wall profile. To validate the numerical results the mean velocity profiles for $\text{Re}_H = 1000$, respectively $\text{Re}_H = 2000$, are compared to measured data obtained by particle image velocimetry (PIV). Figures 8.4 and 8.5 depict the comparison of the mean velocity profiles between PIV and LES at five different positions along the wavy wall for $\text{Re}_H = 1000$, respectively $\text{Re}_H = 2000$. A good agreement between mean experimental data and numerical results is observed. Figure 8.6 depicts the comparison of turbulent kinetic energy profiles obtained by particle image velocimetry and large eddy simulation at two streamwise positions, representing the wave crest ($x/H = 0.00$) and the wave trough ($x/H = 0.50$), at Reynolds numbers $\text{Re}_H = 1000$ and $\text{Re}_H = 2000$. The turbulent kinetic energy from LES is calculated only from the streamwise and the vertical velocity fluctuations in order to resemble the two-dimensionality of the PIV data. It can be seen that for both values of Re_H the profiles exhibit a double peak behavior (in the proximity of the top and bottom walls). For the lower value of $\text{Re}_H = 1000$ the peak value of the turbulent kinetic energy from LES shows higher values and is moved closer to the wall compared to the PIV data. At the identical locations and for the same Reynolds numbers, the vertical profiles of the shear turbulent stress are shown in Figure 8.7. The agreement between the measured and computed profiles in the upper half of the channel ($0.50 \leq y/H \leq 1.00$) is good. However deviations in the magnitude and peak locations are observed in the region above the heated wall, which is most influenced by buoyancy

108   8 Large Eddy Simulation of Mixed Convection

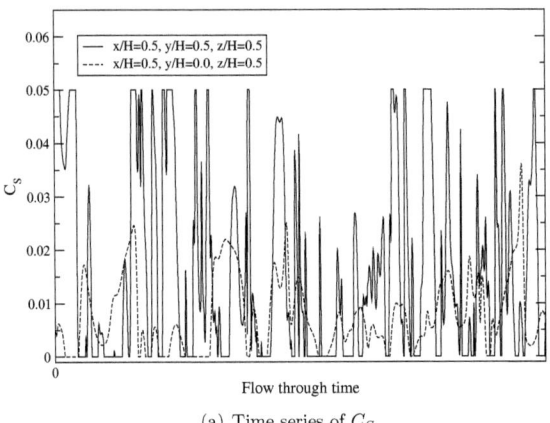

(a) Time series of C_S

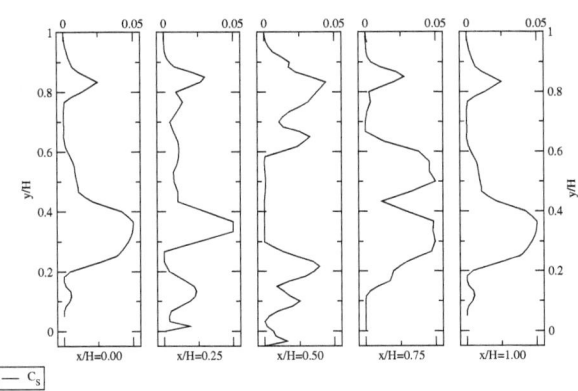

(b) Vertical profiles of C_S

Figure 8.3: Time series and instantaneous vertical profiles of C_S along one wavelength, $\mathrm{Re}_H = 2000$, $\mathrm{Ra}_H = 1.4 \cdot 10^7$.

8.4 Flow Field

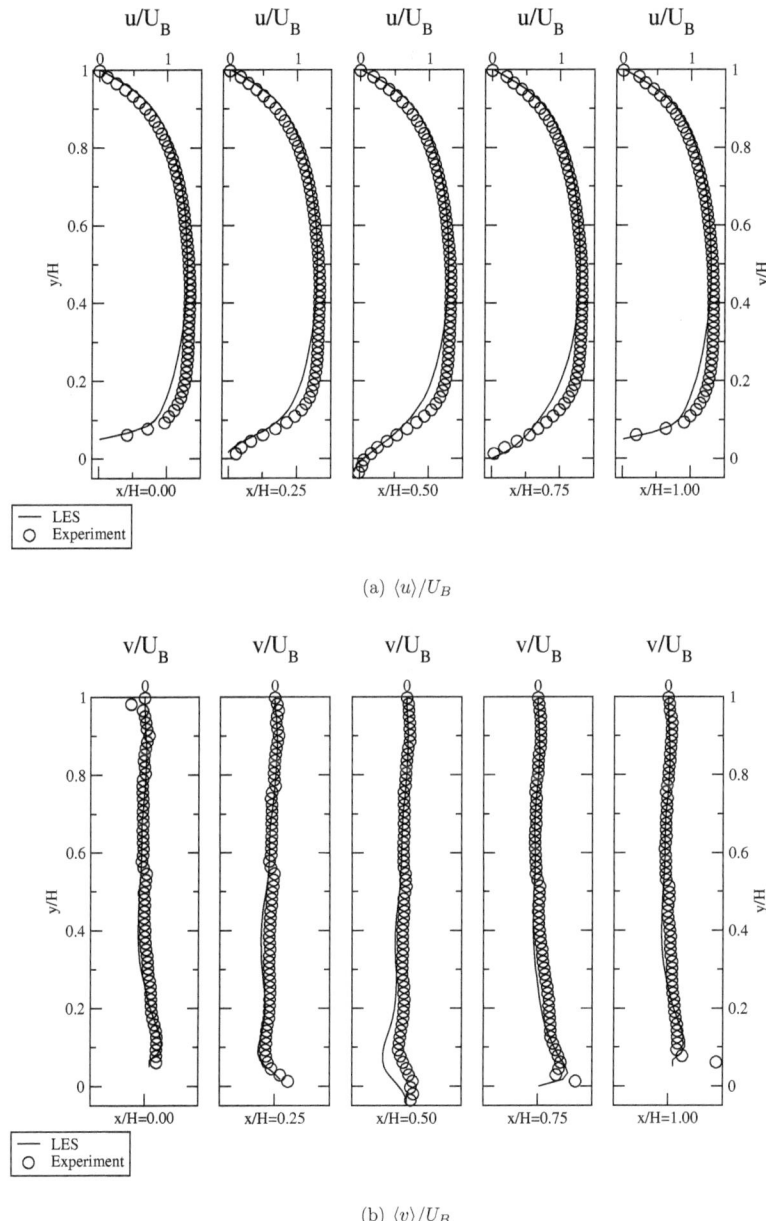

Figure 8.4: Comparison of the mean velocity profiles between particle image velocimetry data and large eddy simulation of mixed convective flow over a wavy wall, $Re_H = 1000$, $Ra_H = 1.4 \cdot 10^7$.

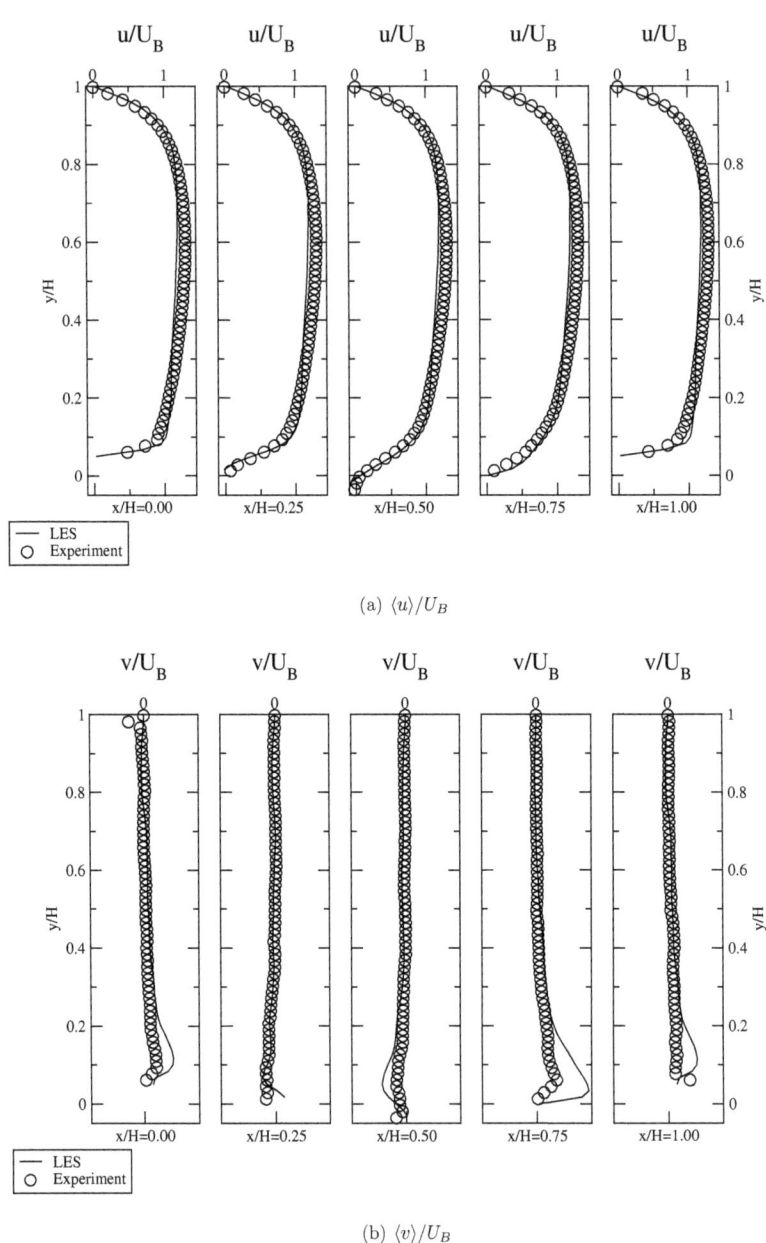

Figure 8.5: Comparison of the mean velocity profiles between particle image velocimetry data and large eddy simulation of mixed convective flow over a wavy wall, $Re_H = 2000$, $Ra_H = 1.4 \cdot 10^7$.

8.4 Flow Field

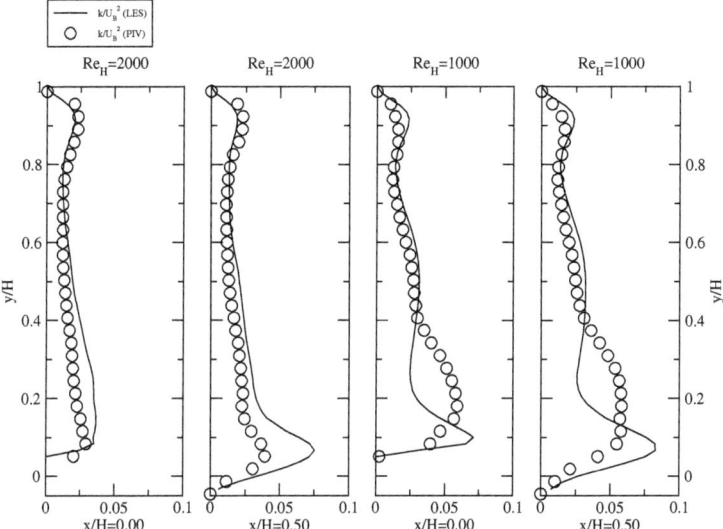

Figure 8.6: Comparison of turbulent kinetic energy profiles between particle image velocimetry data and large eddy simulation of mixed convective flow over a wavy wall at two streamwise positions for $Re_H = 1000$ and $Re_H = 2000$, $Ra_H = 1.4 \cdot 10^7$.

effects. These differences can be explained by a statistically insufficient ensemble size of the PIV measurements, together with measurement uncertainties caused by the changed optical properties of the fluid due to heating.

To address the influence of the strength of the thermal buoyancy (from $Re_H = 200$ to 2000) on the flow reorganization, the mean streamwise velocity profiles are shown in Figure 8.8. Negative values of the velocity at the streamwise position $x/H = 0.50$ found for all values of the Reynolds number indicate flow separation and the presence of a recirculation zone at the downstream side of the wave. For $Re_H = 2000$ the maximum flow velocity is found in the upper part of the channel ($y/H \approx 0.7$) which also results in an asymmetric profile. With decreasing Reynolds number, i.e. increasing impact of the natural convection, the location of the maximum flow velocity is shifted towards the heated surface. For $Re_H = 1000$ the velocity profile is symmetric with the velocity maximum in the center of the channel. By further decreasing the Reynolds number the velocity profile is deformed, and the location of maximum fluid velocity is found at $y/H \approx 0.15$ for $Re_H = 200$.

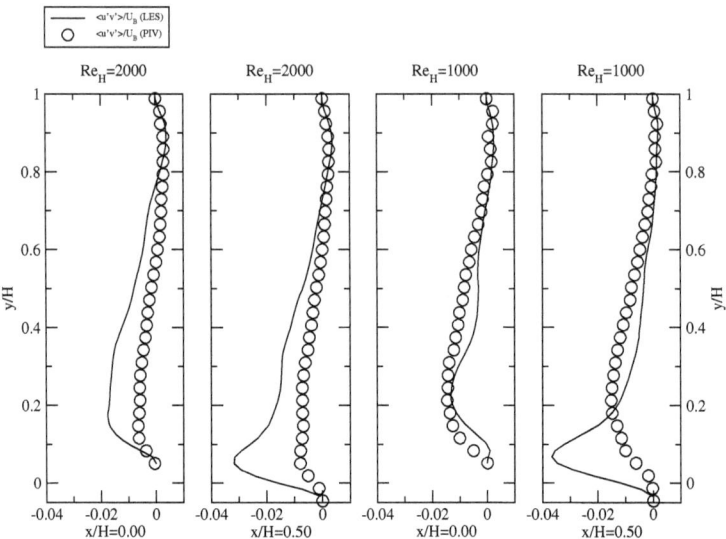

Figure 8.7: Comparison of Reynolds stress profiles between particle image velocimetry data and large eddy simulation of mixed convective flow over a wavy wall at two streamwise positions for $Re_H = 1000$ and $Re_H = 2000$, $Ra_H = 1.4 \cdot 10^7$.

8.4.2 Turbulent Kinetic Energy, Turbulence Anisotropy and Coherent Structures

In order to identify the main production mechanism behind the turbulent mixing over a wavy wall, contours of the resolved turbulent kinetic energy $k = 0.5 \cdot \left(\overline{u'u'} + \overline{v'v'} + \overline{w'w'} \right)$ for different values of $Re_H = 20$, 200, 1000, and 2000 in the central vertical plane $(z/H = 0.5)$ are shown in Figure 8.9. The values are made dimensionless by a characteristic mixed velocity defined as

$$U_m = \sqrt{U_B^2 + U_{buoy}^2}, \qquad (8.14)$$

where the buoyancy velocity U_{buoy} is calculated according to

$$U_{buoy} = \sqrt{\beta g \Delta T H}. \qquad (8.15)$$

It can be seen that for $Re_H = 20$ the maximum of the resolved turbulent kinetic energy is identified in the central part of the domain above the wave crest. In this case, the major contribution of the turbulent kinetic energy comes from the thermal buoyancy production. For the intermediate value of $Re_H = 200$, the two distinct high turbulent kinetic energy regions are moved towards the center of the simulated domain. For higher Reynolds numbers, $Re_H = 1000$ and 2000, the maximum values are observed in the proximity of the bottom wavy wall. Another approach to characterize the influence of the thermal buoyancy on the flow structures is to

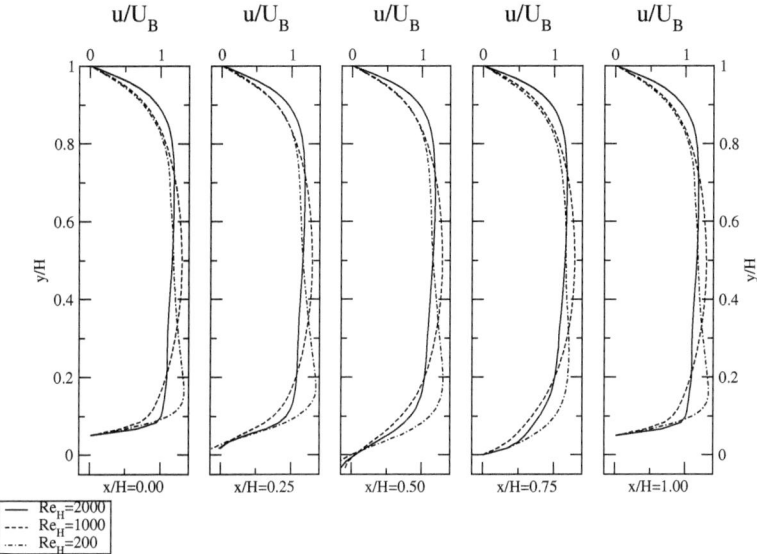

Figure 8.8: Comparison of the normalized mean velocity u/U_B for $\text{Re}_H = 2000$, $\text{Re}_H = 1000$, and $\text{Re}_H = 200$; $\text{Ra}_H = 1.4 \cdot 10^7$.

address the anisotropy of the flow, expressed by the anisotropy flatness parameter A, which is defined according to Lumley and Newman (1977)

$$A = 1 - \frac{9}{8}(A_2 - A_3), \tag{8.16}$$

where

$$A_2 = a_{ij}a_{ji}, \quad A_3 = a_{ij}a_{jk}a_{ki}, \tag{8.17}$$

and

$$a_{ij} = \frac{\overline{u'_i u'_j}}{k} - \frac{2}{3}\delta_{ij}. \tag{8.18}$$

Figure 8.10 depict the contours of the anisotropy parameter A, which equals 1 for isotropic turbulence. It can be observed that this state of isotropic turbulence is reached in the center of the flow channel for Reynolds numbers $\text{Re}_H = 1000$ and $\text{Re}_H = 2000$, similarly to the fully developed channel flow with flat walls. Low values of $A \approx 0$ near the wall for all four cases represent an increase in anisotropy due to the wall influence. For $\text{Re}_H = 20$ a region where $A \approx 0$ is also found just above the wave crest extending nearly over the entire channel height. This high level of the turbulence anisotropy is caused by a convective eddy (roll/cell structure) similar to the standard Rayleigh-Bénard convection situation. In order to visualize these eddy structures for different values of Re_H, the swirling strength criteria according to Zhou et al.

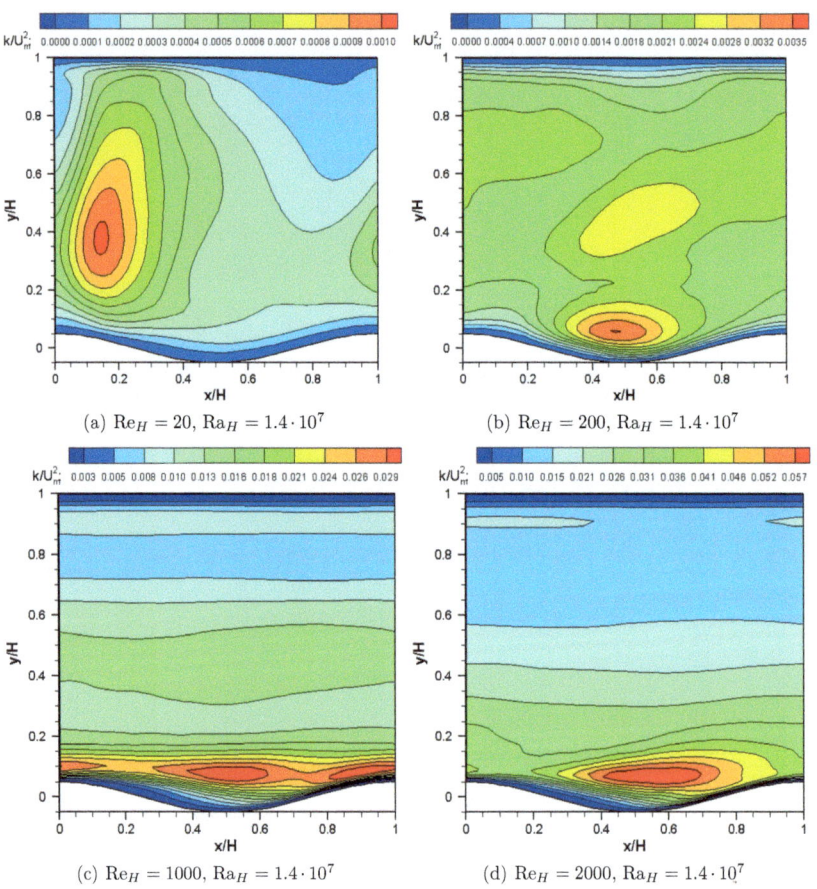

Figure 8.9: Contours of turbulent kinetic energy k/U_m^2 for $\mathrm{Re}_H = 2000$, $\mathrm{Re}_H = 1000$, $\mathrm{Re}_H = 200$, and $\mathrm{Re}_H = 20$; $\mathrm{Ra}_H = 1.4 \cdot 10^7$.

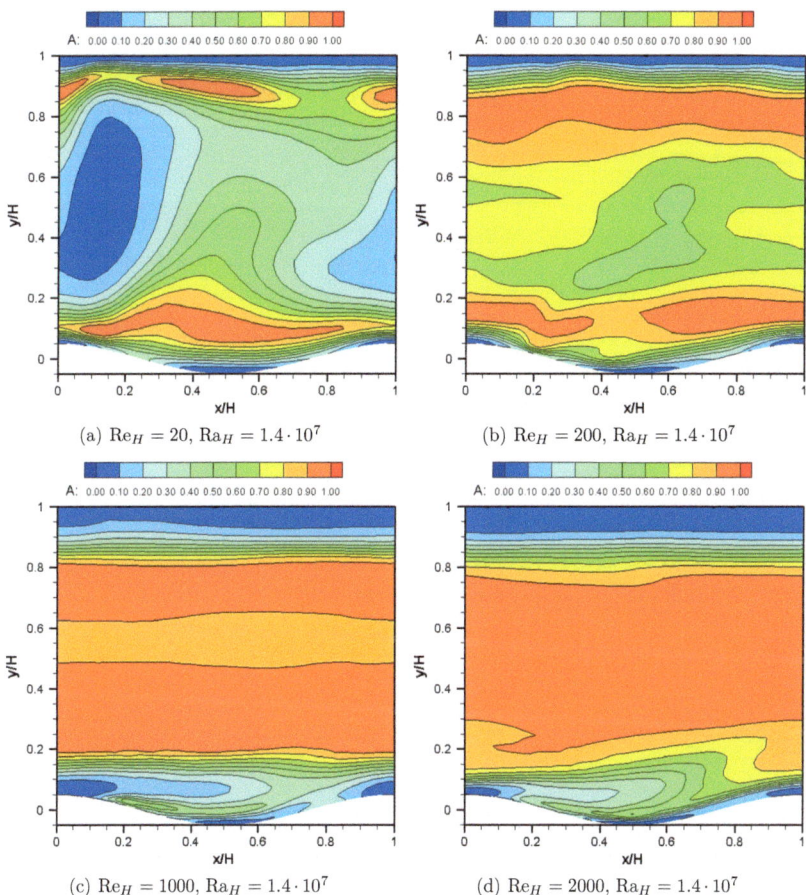

Figure 8.10: Contours of the anisotropy parameter A for $\mathrm{Re}_H = 2000$, $\mathrm{Re}_H = 1000$, $\mathrm{Re}_H = 200$, and $\mathrm{Re}_H = 20$; $\mathrm{Ra}_H = 1.4 \cdot 10^7$.

(1999) and Adrian et al. (2000) is used for their identification and extraction. The swirling strength λ_{ci} is defined as the imaginary part of the complex eigenvalues of the local velocity gradient tensor. In a three–dimensional flow field the local velocity gradient tensor will have one real eigenvalue and a pair of complex conjugate eigenvalues when the discriminant of its characteristic equation is positive. When this is the case, the particle trajectories about the eigenvector corresponding to the real eigenvalue exhibit a spiral, swirling motion, as shown in Chong et al. (1990).

Figure 8.11 depicts the isosurfaces of constant swirling strength λ_{ci} for all Reynolds numbers. It is important to observe that these regions with strong swirling motion coincide very closely to the regions where the levels of turbulent anisotropy were highest. For the lowest value of $Re_H = 20$, the eddy structures are oriented with a central axis perpendicular to the mean flow and resemble closely convective Rayleigh-Bénard cells (Hanjalić and Kenjereš (2002)). With further increase of the Reynolds number to $Re_H = 200$, these eddy structures start to reorganize in a way that their central axis is now aligned with the flow in the streamwise direction. For the two highest values of $Re_H = 1000, 2000$, these large cell structures disappear and regions with intensive swirling can be observed only in the proximity of the wavy wall, thus portraying separation and recirculation zones (Figure 8.11(c-d)).

8.5 Scalar Field

In this section we address the distribution of temperature as an active scalar and the influence of thermal buoyancy on the wall heat transfer.

8.5.1 Mean Temperature and Temperature Variance

The vertical profiles of the mean temperature and temperature variance (normalized by a reference temperature) for $Re_H = 1000, 2000$ at two characteristic locations ($x/H = 0, 0.5$) are shown in Figure 8.12. The LES results show generally good agreement with LIF for the mean temperature profiles for both values of Re_H at both locations. The only exception is the location $x/H = 0.5$ for $Re_H = 2000$, where the measured wall temperature is higher compared to the LES value. The temperature variance profiles show a good agreement in the proximity of the wavy wall, but significant deviations are observed in the proximity of the upper wall (Figure 8.12(b)). Here, the LES profiles exhibit a proper near–wall behavior since for the applied constant temperature boundary condition, the temperature variance should have a zero boundary value. In contrast to that, the measured profiles show a finite value at the wall indicating heat losses through this boundary during the experiment.

Contours of the mean temperature and temperature variance in the central vertical plane for different values of Re_H are given in Figures 8.13 to 8.16. In addition, the streamlines are superimposed on the mean temperature field in order to illustrate the mutual dependency between the underlying velocity field with temperature distributions. It can be seen that the temperature variance reaches its maximum at the separation point. The locations of the separation and reattachment points for different values of Re_H are given in Table 8.2. The separation point moves from $x_S/H = 0.1$ to 0.16 for $Re_H = 20$ and 2000, respectively. For the same conditions, the reattachment point moves from $x_R/H = 0.54$ to 0.66. This behavior indicates a horizontal shift and a vertical reduction of the separation bubble. This is a consequence of the increasing forced convection influence over thermal buoyancy. It is interesting to compare

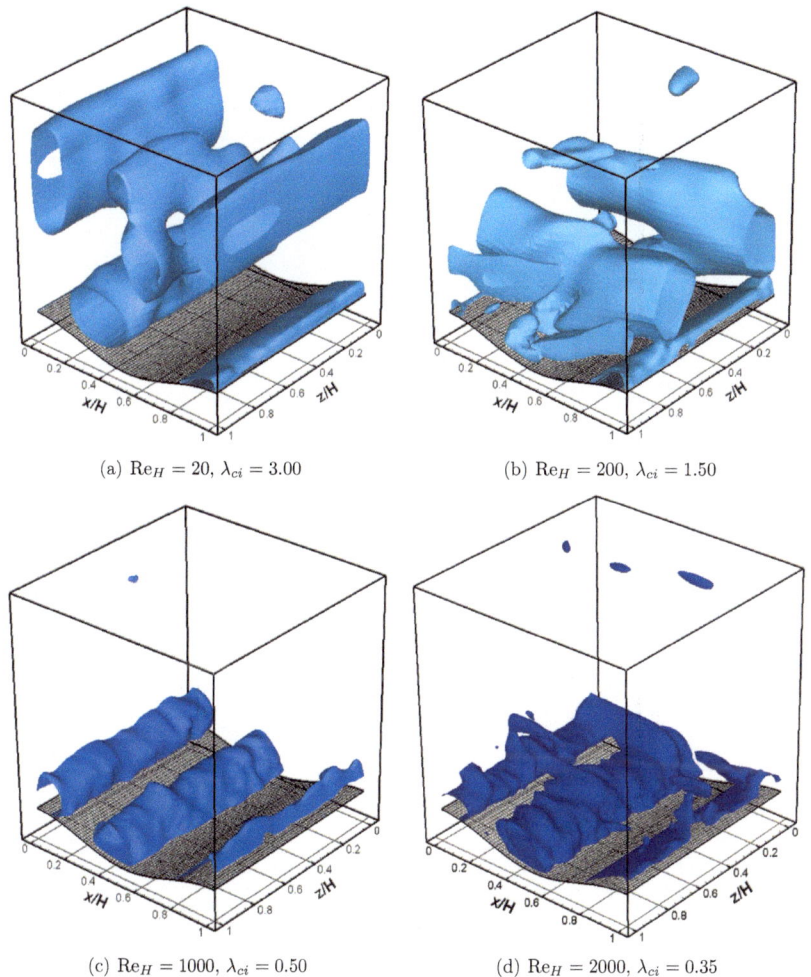

Figure 8.11: Isosurfaces of constant swirling strength λ_{ci} for $\mathrm{Re}_H = 2000$, $\mathrm{Re}_H = 1000$, $\mathrm{Re}_H = 200$, and $\mathrm{Re}_H = 20$; $\mathrm{Ra}_H = 1.4 \cdot 10^7$.

118 8 Large Eddy Simulation of Mixed Convection

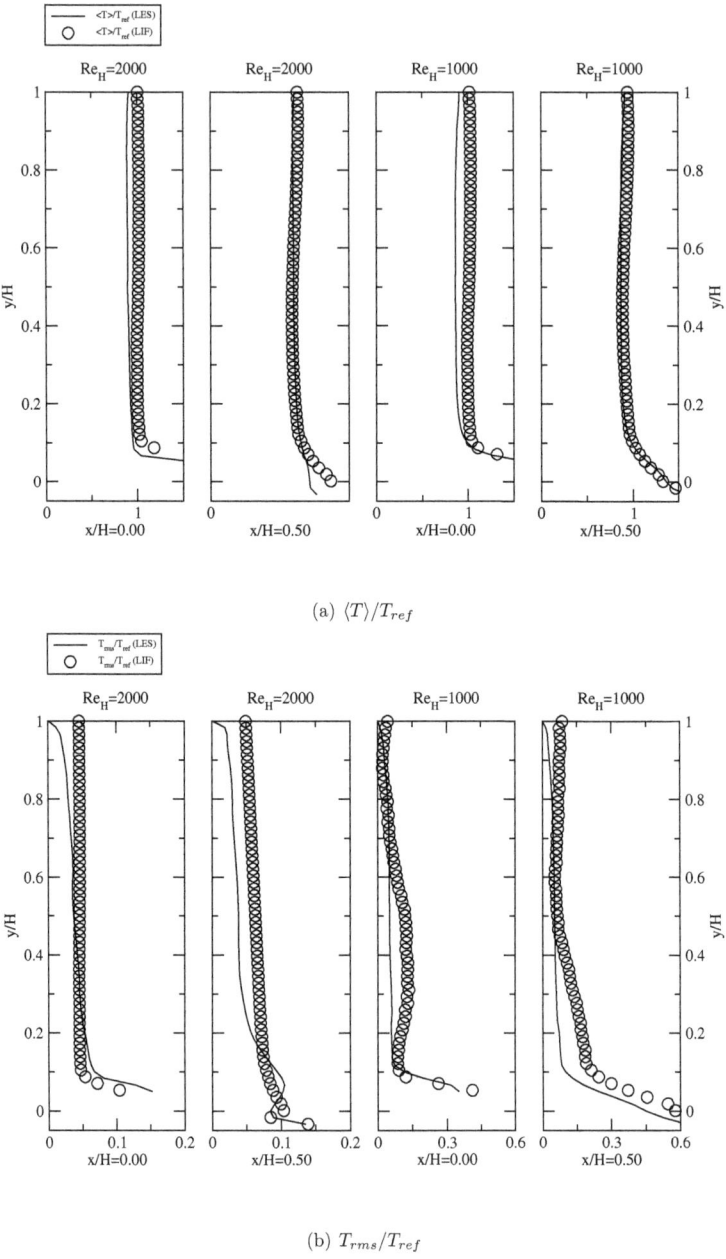

Figure 8.12: Comparison of the mean temperature profiles and the temperature fluctuations between laser induced fluorescence and large eddy simulation.

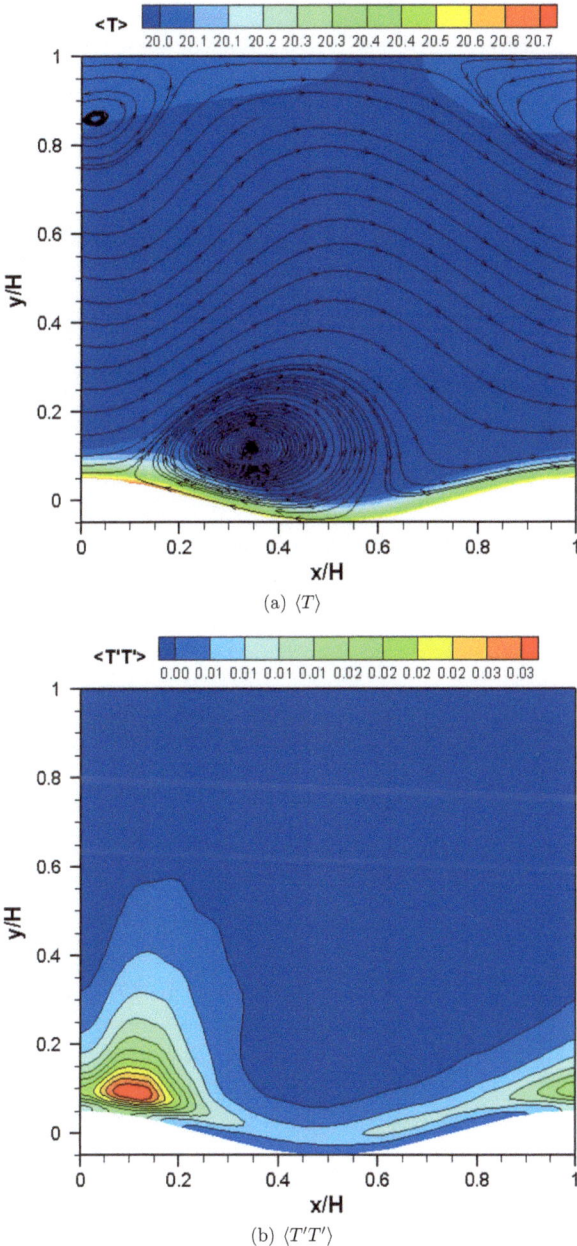

Figure 8.13: Mean temperature field and temperature variance together with streamlines, $\mathrm{Re}_H = 20$, $\mathrm{Ra}_H = 1.4 \cdot 10^7$.

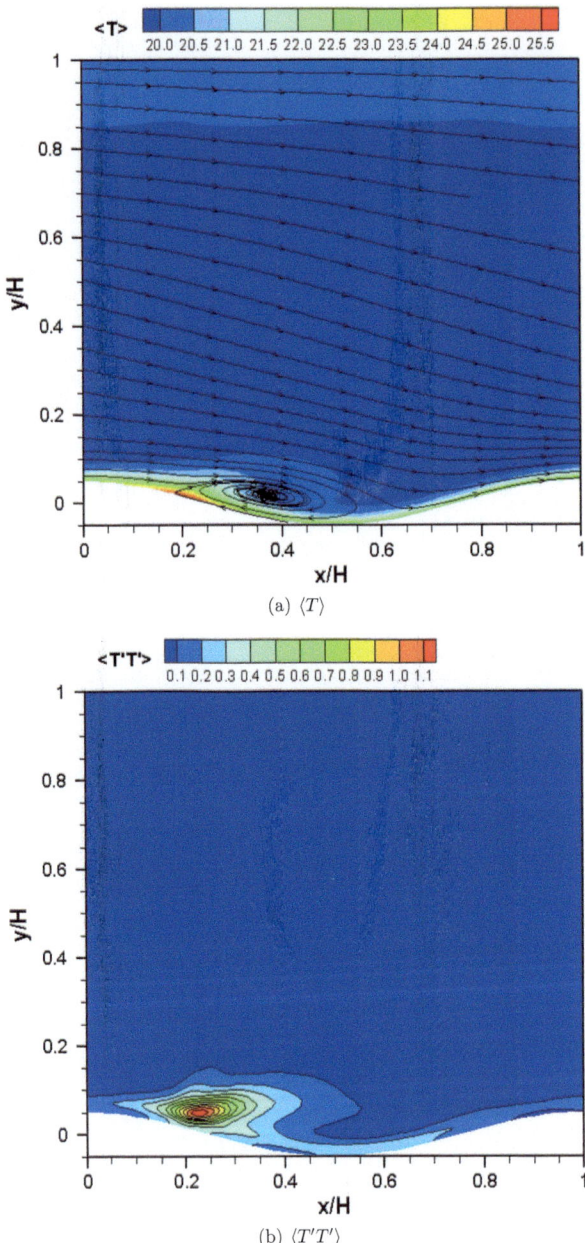

Figure 8.14: Mean temperature field and temperature variance together with streamlines, $\mathrm{Re}_H = 200$, $\mathrm{Ra}_H = 1.4 \cdot 10^7$.

8.5 Scalar Field

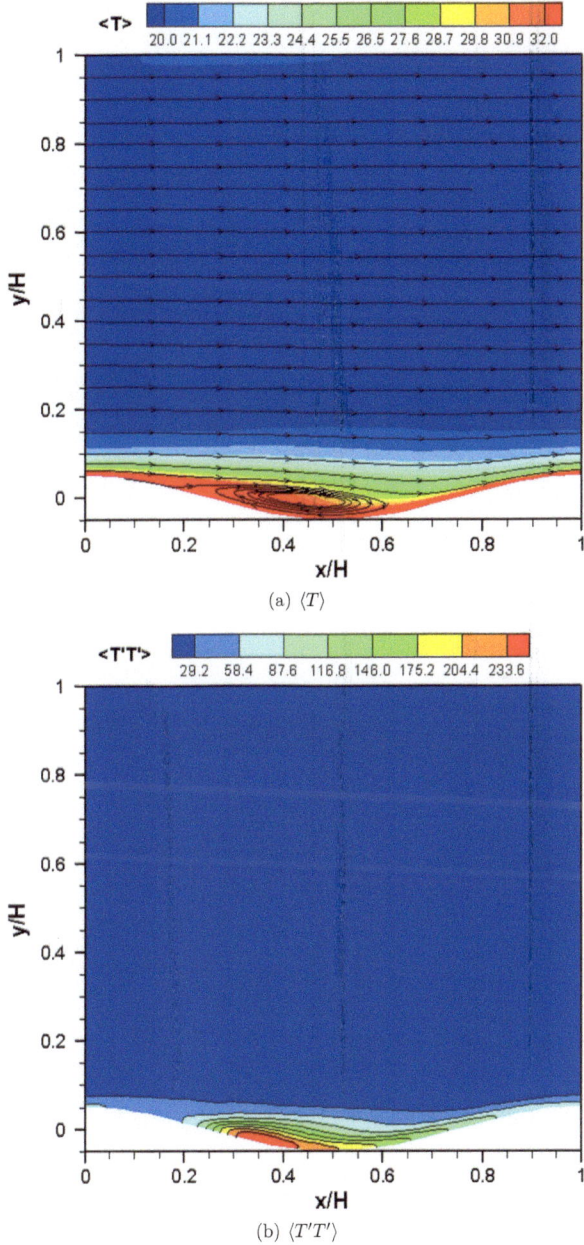

(a) $\langle T \rangle$

(b) $\langle T'T' \rangle$

Figure 8.15: Mean temperature field and temperature variance together with streamlines, $\text{Re}_H = 1000$, $\text{Ra}_H = 1.4 \cdot 10^7$.

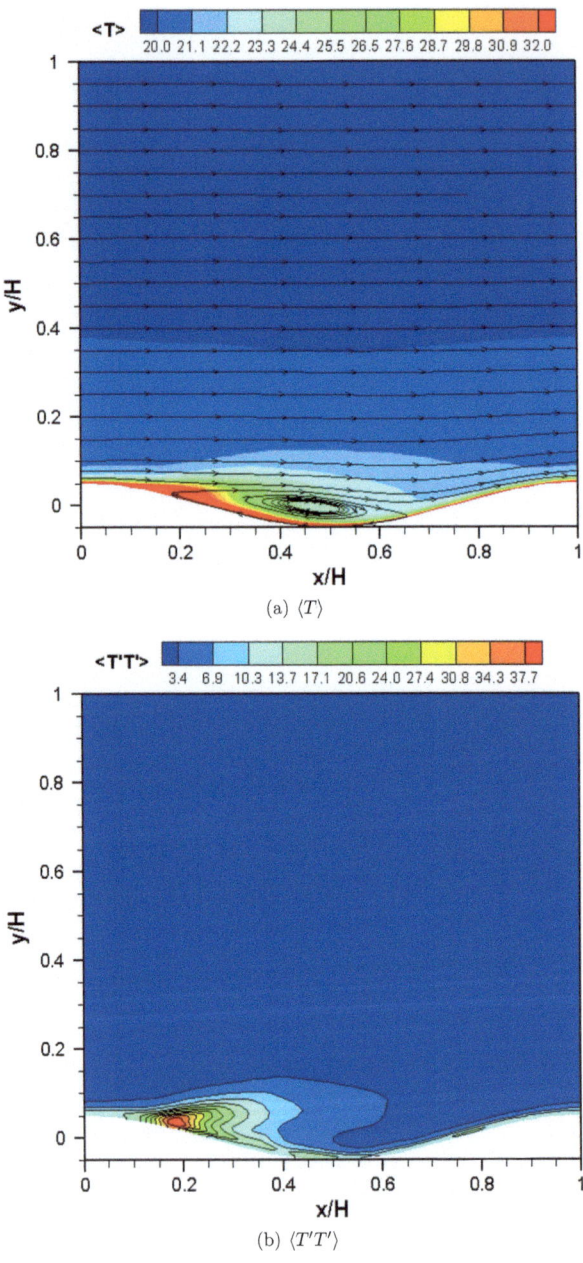

Figure 8.16: Mean temperature field and temperature variance together with streamlines, $\mathrm{Re}_H = 2000$, $\mathrm{Ra}_H = 1.4 \cdot 10^7$.

Table 8.2: Streamwise coordinate of the separation x_S/H and reattachment x_R/H points for the mixed convective flow over waves.

Re_H	Ra_H	Ri	x_S/H	x_R/H
20	$1.4 \cdot 10^7$	5000	0.10	0.54
200	$1.4 \cdot 10^7$	50	0.18	0.60
1000	$1.4 \cdot 10^7$	2	0.18	0.65
2000	$1.4 \cdot 10^7$	0.5	0.16	0.66

these locations of separation and reattachment with a thermally neutral situation (no heating). In such a case, the separation points are located at $x_S = 0.24, 0.19$ and the reattachment points at $x_R/H = 0.65, 0.72$ for $Re_H = 1000, 2000$, respectively. These values obviously differ from the thermally active situations indicating that thermal buoyancy effects still play an important role for these two highest simulated values of Re_H.

8.5.2 Turbulent Heat Flux and Wall Heat Transfer

Figure 8.17 depicts the comparison of the heat flux profiles between experimental and numerical data at two streamwise positions for $Re_H = 1000$ and $Re_H = 2000$. As already stated for the velocity fluctuations, the agreement between the measured and computed profiles in the upper half of the channel ($0.50 \leq y/H \leq 1.00$) is good for $Re_H = 2000$. However, deviations in the magnitude and peak locations are observed in the region above the heated wall, where the thermal buoyancy is the most dominant. These differences between simulations and experiments can be explained by a statistically insufficient ensemble size of the PIV/LIF statistics together with measurement uncertainties caused by the changed optical properties of the fluid due to heating. Contours of the horizontal and vertical turbulent heat flux components are shown in Figures 8.18 to 8.21. For the case $Re_H = 20$ the vertical component of the turbulent heat flux vector is significantly larger in comparison to the horizontal component. It is interesting to observe that, for this situation, the highest value of the vertical turbulent heat flux does not coincide with the location of the strong recirculation but is concentrated in a region bordered by the already identified and discussed coherent structures (Figure 8.11). In contrast to this situation, for higher values of Re_H, the horizontal heat flux component takes over the vertical component. The highest values of the vertical turbulent flux are concentrated in the region above the recirculation zones (Figure 8.21).

The wall heat transfer, characterized by a local Nusselt number distribution, is presented next. We define the Nusselt number as ratio between the total heat flux and the conductive heat flux, which reads

$$Nu = \frac{\dot{q}H}{\lambda(T - T_{ref})} + 1. \tag{8.19}$$

Figure 8.22 depicts the contours of the Nusselt number at the heated surface for all Reynolds numbers ($Ra_H = 1.4 \cdot 10^7$). The wave-averaged Nusselt numbers for each case are tabulated in Table 8.3. In general, an increase of the Nusselt number with increasing Reynolds number is observed. However, an interesting transition occurs between $Re_H = 200$ and 1000. It can be seen that the averaged value of the Nusselt number is higher for the smaller value of Re_H. This can be explained by an active role of the coherent structures in transporting mass, momentum

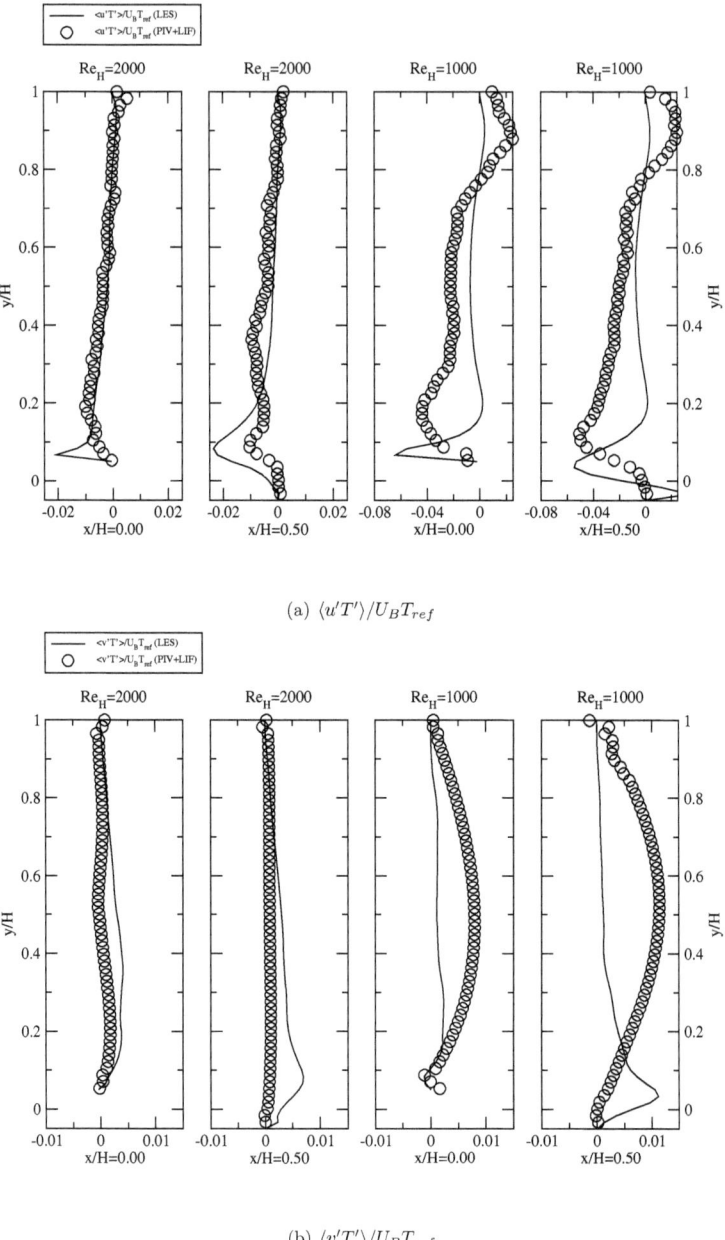

Figure 8.17: Comparison of the heat flux profiles between combined particle image velocimetry and laser induced fluorescence data and large eddy simulation.

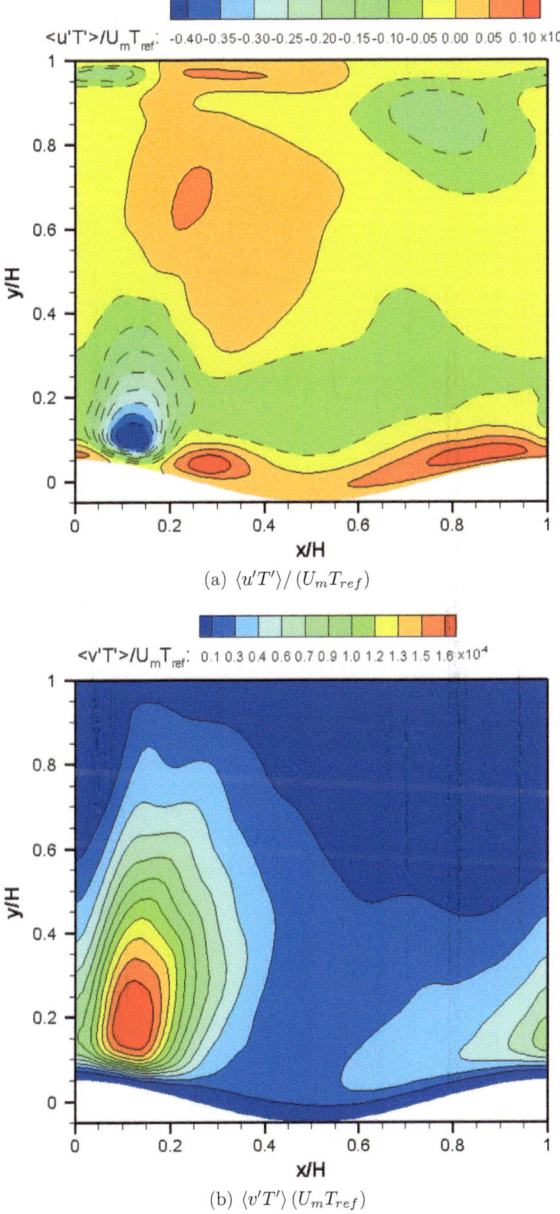

Figure 8.18: Contours of streamwise and vertical heat flux, $\mathrm{Re}_H = 20$, $\mathrm{Ra}_H = 1.4 \cdot 10^7$.

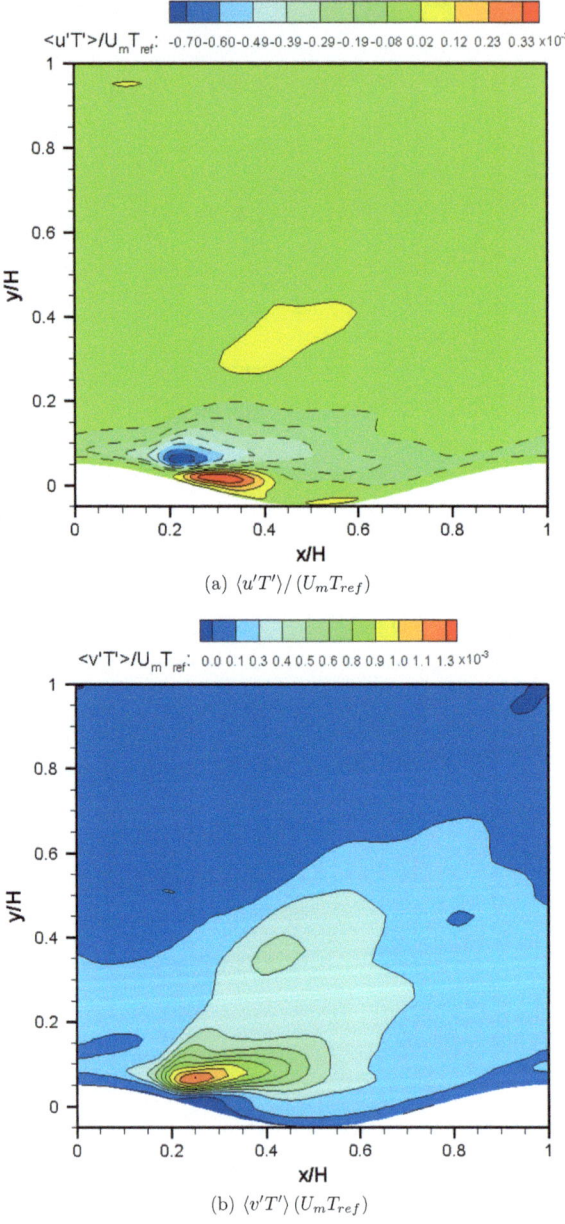

Figure 8.19: Contours of streamwise and vertical heat flux, $Re_H = 200$, $Ra_H = 1.4 \cdot 10^7$.

8.5 Scalar Field

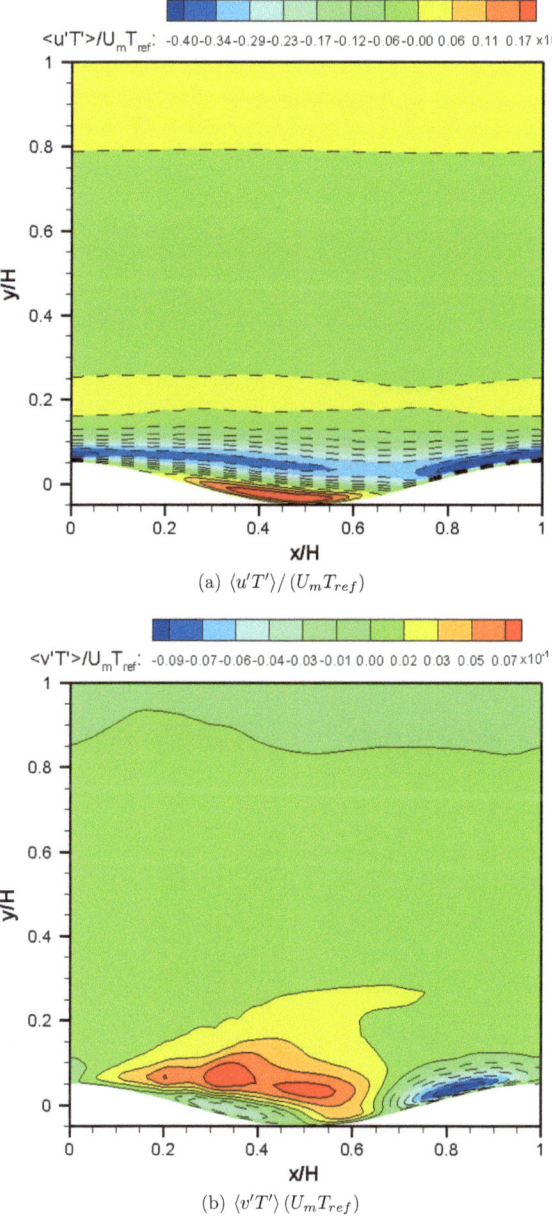

Figure 8.20: Contours of streamwise and vertical heat flux, $\text{Re}_H = 1000$, $\text{Ra}_H = 1.4 \cdot 10^7$.

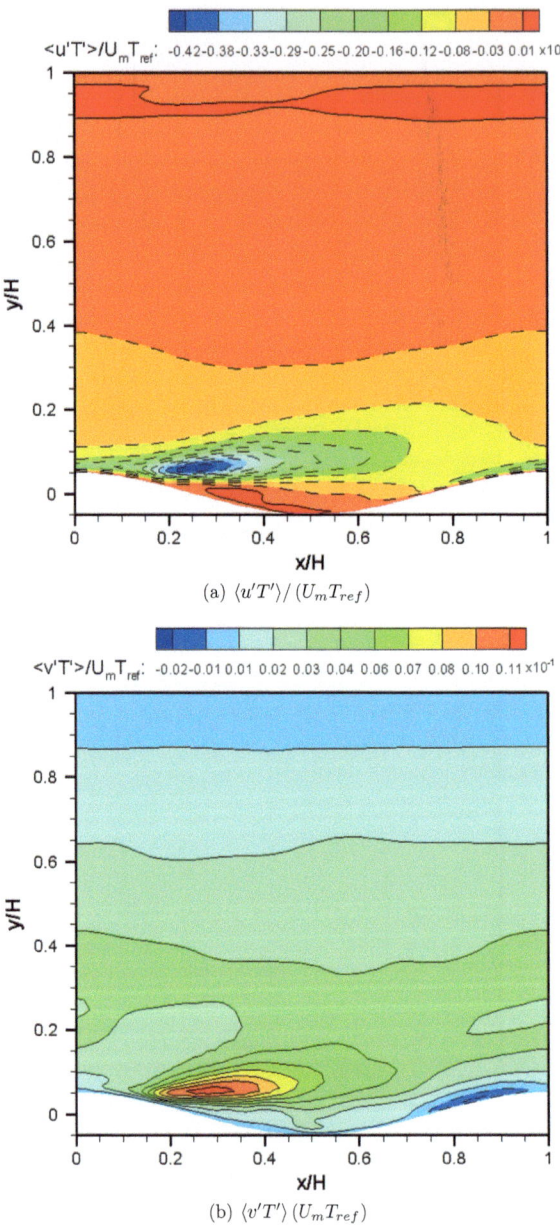

Figure 8.21: Contours of streamwise and vertical heat flux, $Re_H = 2000$, $Ra_H = 1.4 \cdot 10^7$.

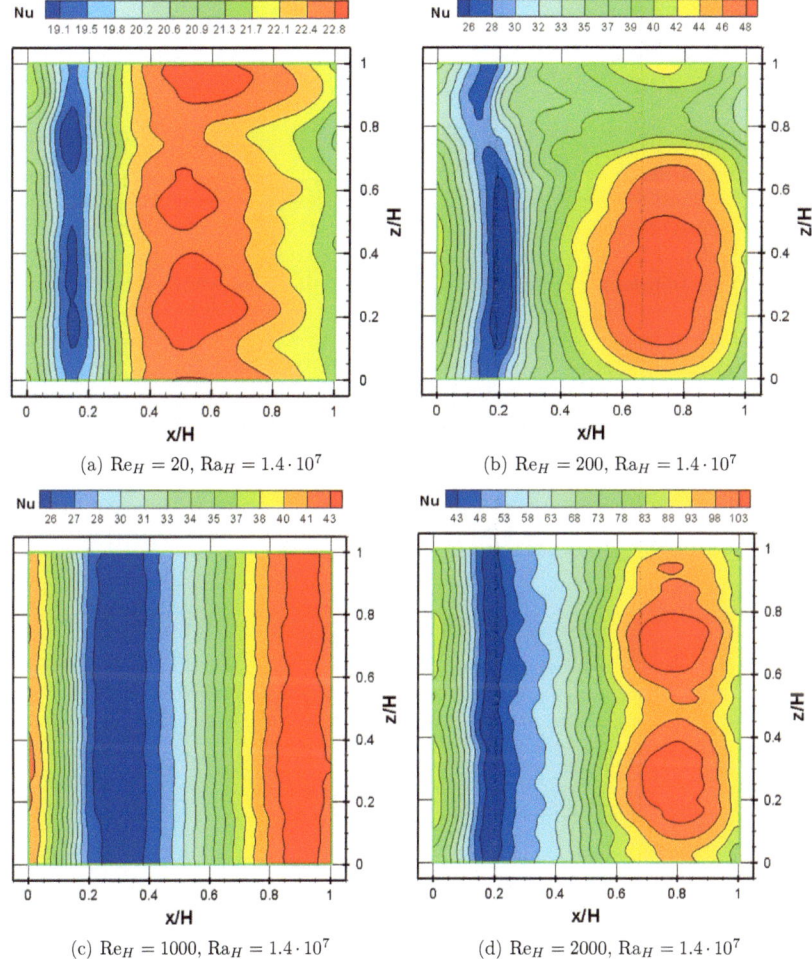

Figure 8.22: Contours of the Nusselt number Nu for $Re_H = 2000$, $Re_H = 1000$, $Re_H = 200$, and $Re_H = 20$; $Ra_H = 1.4 \cdot 10^7$.

Table 8.3: Comparison of mean Nusselt numbers with experiments and predictions for forced and free convection.

Re_H	Ra_H	Nu_{LES}^{mixed}	Nu_{exp}^{mixed}	Nu_{exp}^{forced}	Nu_{exp}^{free}
20	$1.4 \cdot 10^7$	21.70	–	–	17.75
200	$1.4 \cdot 10^7$	39.03	–	–	–
1000	$1.4 \cdot 10^7$	34.02	36.00	15.29	–
2000	$1.4 \cdot 10^7$	75.37	90.00	28.05	–

and heat. For the smaller value of Re_H, these structures extend over the entire domain, i.e. they are not confined to the recirculation areas only, as shown in Figure 8.11. In addition to the heat transfer enhancement, these coherent structures also cause a strong non–uniformity of the wall heat transfer. This can be easily observed by comparing the imprints of these structures on the local heat transfer in Figures 8.22(b) and 8.22(c). It can be seen that for the lower value of $Re_H = 200$ a strong asymmetry in the local heat transfer distribution is present.

The integral heat transfer is analyzed next. Here we performed a comparative assessment of numerically obtained results with experimental data. The experimental data include the mixed convection measurements over wavy walls, as well as the experiments performed in purely forced (Gnielinski (1976)), or free convection flow regimes (Chu and Goldstein (1973)). It is important to mention that these experimental studies in purely free or forced convection were performed in configurations with a flat horizontal wall. By comparing our numerical results with these experimental studies, some important conclusions about the effects of the imposed wavy wall on the integral heat transfer can be made. Compared with the experimental correlation of Chu and Goldstein (1973)

$$Nu_{exp}^{free} = 0.183 Ra_H^{0.278}, \tag{8.20}$$

the LES results show a good agreement. The observed difference of 20% heat transfer enhancement is due to the wall waviness. In contrast to that, compared with the experimental correlation for pure forced convection (where the thermal buoyancy effects are neglected) of Gnielinski (1976)

$$Nu_{exp}^{forced} = \frac{0.037 Re_H^{0.8} \Pr}{1 + 2.443 Re_H^{-0.1} \left(\Pr^{2/3} - 1 \right)}, \tag{8.21}$$

the LES results show a significant heat transfer enhancement of 2.2 and 2.68 times for $Re_H = 1000$ and 2000, respectively. Finally, when compared to experiments in identical conditions (mixed convection and wavy horizontal wall, Nu_{exp}^{mixed}) the LES results show a good agreement. The difference for $Re_H = 1000$ is just about 2% while for the higher value of $Re_H = 2000$ an under–prediction of 16 % is obtained. This difference in the Nusselt number can be explained by the averaging procedure. The Nusselt number in LES is obtained by averaging over the whole simulated wavy surface (3D), i.e. the local patterns of the Nusselt number are taken into account. Whereas the experimentally obtained Nusselt number is averaged along the line of measurement (2D), i.e. the patterns of the Nusselt number caused by the flow organization are not taken into account and thus the result differs from the non–homogeneous numerical solution.

8.6 Conclusions

In this chapter we addressed the mixed convective flow over a heated wavy surface by means of a large eddy simulation. The simulations covered Richardson numbers ranging from $Ri = 0.5$ to $Ri = 5000$. Both, first and second order resolved velocity statistics show a good agreement with experimental data.

With increasing influence of natural convection the mean velocity profiles are shifted towards the heated surface. The vortical flow structures are identified and visualized by applying the swirling intensity criteria. An interesting spatial flow reorganization takes place between $Re_H = 20$ and 200, when spanwise oriented rolls start to be aligned in the streamwise direction. Next to causing a significant turbulence anisotropy in the central part of the considered flow domain, the wall imprints of these vortical structures also can be identified in the distributions of the local heat transfer. The predicted values of the integral Nusselt number show a good agreement with recent experimental studies of mixed convection for $Re_H = 1000, 2000$. A good agreement is also obtained for a pure free convection experimental correlation for the lower value of the Reynolds number $Re_H = 20$. Compared to a standard experimental Nusselt number correlation for forced convection over a flat wall, significant enhancements in heat transfer have been obtained by a factor of 2.2 and 2.68 for $Re_H = 1000$ and 2000, respectively. Thus we have been able to demonstrate the wall heat transfer enhancement in mixed convective flows and the reorganization of the flow due to buoyancy effects.

Chapter 9
Conclusions

In this thesis the transport properties of flows in the forced convective and in the mixed convective regime over different wavy walls have been addressed. Therefore we employed a combined measurement technique of particle image velocimetry and laser induced fluorescence to simultaneously assess the fluid velocity field and the scalar field, i.e. the fluid temperature and the concentration of a species. In addition, we numerically predicted the mixed convective flow over a wavy surface by means of large eddy simulation. For the identification of coherent structures we applied the proper orthogonal decomposition and the criteria of constant swirling strength. In the following paragraphs we shortly summarize the main findings:

Flows in the forced convective regime From the orthogonal decomposition of velocity fields we find similar large–scale structures in the vicinity of the complex surface. These large–scale structures are identified as streamwise–oriented, counter–rotating vortices exhibiting a characteristic scale in the spanwise coordinate direction. This scale is found to be sensitive to alterations in the wavy surfaces. For the three basic two–dimensional wave profiles a similar spanwise scale in the order of $\Lambda_z = 1.5H$ is identified for the first two eigenmodes, which reduces to $\Lambda_z = 0.85H$ in the first mode and $\Lambda_z = 1.3H$ in the second mode for the three–dimensional surface. In addition, the eigenvalue spectra become increasingly broader for the profile with doubled amplitude ($\alpha = 0.2$ ($\Lambda = 30$ mm)), half the wavelength ($\alpha = 0.2$ ($\Lambda = 15$ mm)), and the three–dimensional superimposed waves. Thus we concluded that by altering, i.e. increasing the surface complexity more eigenmodes contribute to the energy containing range. In addition, we find a significant impact of these coherent structures on the transport of a species when comparing two- and three–dimensional surfaces. In contrast to a free turbulent jet the influence of the structured surfaces enhances the scalar transport normal to the mean flow direction. By comparing the two- and three–dimensional wavy surfaces it is found that the superimposed waves promote the scalar transport in spanwise and vertical direction compared to the basic sinusoidal surface. Due to the flow inhomogeneity in all three coordinate directions and the smaller spanwise scale of the large–scale structures the mixing properties of the turbulent flow are significantly enhanced.

Flows in the mixed convective regime Advancing to flows in the mixed convective regime we observe several changes in the flow physics. With increasing effect of buoyancy induced motions the mean velocity profile shifts towards the heated surface. In addition, flow separation occurs earlier compared to the isothermal case. This coupling between heat as an active scalar and the flow itself is also observed in the turbulence quantities. We conclude that momentum transport is additionally increased in the mixed convective regime. The maximum of the heat flux coincides with the maximum of the Reynolds shear stress, respectively with the spatial orientation of large–scale structures identified by POD, which means that locations of increased momentum transport coincide with the regions of increased scalar transport. Thus we have been able to show a correlation of momentum and scalar transport in mixed convective flow over waves. Addressing the transport of a species introduced by a point source we find enhanced vertical transport due to buoyancy effects and enhanced lateral transport due to the presence of the afore mentioned large-scale structures. These experimental results are supported by the numerical study. In addition, the result of the simulations provide more insights into the heat transfer from the wavy surface. The patterns of the Nusselt number are strongly connected to the spatial organization of coherent structures. The comparison of the obtained Nusselt numbers with theoretical predictions yields larger values for the mixed convection, which supports our statement that transport properties are increased by buoyancy induced fluid motions. Insight into this fact is given when calculating the anisotropy of the flow, in mixed convection the turbulence state in the center of the flow channel is more anisotropic than in classical channel flow, i.e. transport normal to the mean flow is favored.

We identified coherent structures in the flow field which are directly linked to the transport mechanisms. These coherent structures appear to be sensitive to alterations in the structure of the wavy surfaces, allowing some sort of turbulence control, i.e influencing transport properties through the structure of the bounding walls. However, we have to state that the generation process of these coherent structures was not studied in this thesis. Therefore a detailed investigation of the near wall region of the flow should be conducted. In the mixed convection the flow state in the domain becomes more anisotropic, resulting in increased transport of momentum and scalars. However, the strong connection between momentum transport (i.e. large-scale structures) and active scalar transport is also present in this case. Thus all our findings support the notion that transport mechanisms important in complex flow conditions are governed by large-scale structures.

Chapter 10
Outlook

This section tries to give directions for future research with respect to applications, and the use of new measurement and numerical techniques.

10.1 Reactive Flows

An additional increase in complexity of the involved transport mechanisms is the transition from the passive mixing processes studied in this thesis to active scalar mixing introduced by a chemical reaction. A simultaneous measurement technique combining particle image velocimetry and laser induced fluorescence with pH-value dependant dyes can be set up to investigate the momentum field and to capture the front of the ongoing chemical reaction. Coppeta and Rogers (1998) investigated different dyes and identified several candidates which are excitable with an Ar–Ion laser (either at λ_{ex} = 488 nm or λ_{ex} = 514 nm) and exhibit a dependency of their emitted light on the pH in the range between pH = 6 and pH = 9. With such a dye system it is feasible to investigate a neutralization between an acid and a base. Komori et al. (1991) and Komori et al. (1993) investigated the neutralization of acetic acid with ammonium hydroxide in a channel flow facility employing point–measurement techniques (LDV and LIF) to record the velocity and the concentration of acetic acid. In their experimental setup the two reacting streams are mixed via a grid generated turbulence. Nagata and Komori (2000) investigated the effects of unstable thermal stratification (i.e. buoyancy effects) and mean shear on a chemical reaction in grid turbulence. Their main finding was that the buoyancy induced motions promote mixing processes and the reaction was much more efficient compared to shear. Mitrovic and Papavassiliou (2004) conducted a numerical simulation to investigate the effect of a chemical reaction on mass transfer in channel flow. In their computational setup one of the reacting species was introduced via a line source in the bottom wall. With modifications to the existing flow facility similar experiments could be conducted, where the simultaneous measurement of the velocity field and of the reaction in a whole channel cross–section would reveal more insights into active mixing processes and could be compared to numerical predictions.

10.2 Measurement Techniques

10.2.1 Tomographic PIV

Recent developments in the field of PIV led to a three–dimensional measurement system which is known as tomographic particle image velocimetry. The basics of tomographic PIV are described in detail by Elsinga et al. (2006), and are introduced briefly in the following. In contrast to ordinary PIV the laser light is not expanded to a thin light sheet but illuminates a whole volume of the flow. The scattered light from the seeding particles is recorded by four individual CCD cameras. Through a tomographic reconstruction of the volume and a subsequent cross-correlation the velocity vector field in the whole volume is obtained. Schröder et al. (2008) applied time–resolved tomographic PIV to a turbulent boundary layer, thus proving the feasibility to apply this measurement technique to relevant flow cases. The application of tomographic PIV to the flow over waves could provide additional experimental data to address the spatial organization of large–scale structures and to compare with numerical results.

10.2.2 Multiphase flows

For applications in multiphase flows the simultaneous measurement of the velocity of each phase is of interest. In a liquid-liquid system this is achieved by applying PIV with two cameras simultaneously. The first phase is seeded with ordinary seeding particles scattering light with $\lambda_{scat} = 532$ nm, the second phase with Rhodamine B coated particles scattering light with $\lambda_{scat} = 580$ nm. By applying filters to each camera the signals can be separated from each other and the cross-correlation yields the individual velocity of each phase. This information is combined for the whole field of view in the post-processing by using an appropriate mapping function.

10.3 Hybrid Modeling Techniques

Having built up a database with large eddy simulations and experiments the next step in the numerical point of view is the transition to less computationally expensive simulations. Hybrid techniques, where the near wall region of the flow is treated in a RANS–like manner, and the outer region of the flow in a LES–like manner, promise a significant reduction in computational time and effort. One hybrid technique is the detached eddy simulation (DES), which uses the Spalart–Allmaras model proposed by Spalart and Allmaras (1994) with a modified distance function, which causes the model to switch from RANS near the wall to LES away from the wall. Since the Spalart–Allmaras belongs to the class of one-equation turbulence models it would be desirable to combine more sophisticated turbulence models with LES. Kenjereš and Hanjalić (2006) report on an approach where LES is combined with a three-equation RANS model (k–ε–θ^2) to predict thermal convection at high Rayleigh numbers in a range of $10^7 \leq \mathrm{Ra} \leq 10^9$. Their model includes seamless coupling between both models, i.e. the switching between the models is not based on geometric considerations but on the flow properties itself. This hybrid approach could be applied to the mixed convective flow over waves and other applications of engineering interest.

Appendix A
Principles of Large Eddy Simulations

A.1 Filtering Operations in LES

The filtering operation in LES is used to decompose a quantity $f(\mathbf{x},t)$ into the sum of a filtered (or resolved) component $\langle f \rangle(\mathbf{x},t)$, and the residual (or unresolved) component $f'(\mathbf{x},t)$. This spatially filtered values are obtained by applying a filter kernel $G(x,x')$, i.e. a function which is large when x' is close to x, and is small elsewhere. The general filtering operation is described by

$$\langle f \rangle = \int G(x,x') f(x') \mathrm{d}x', \tag{A.1}$$

where the filter kernel satisfies the normalization condition

$$\int G(x,x') \mathrm{d}x' = 1. \tag{A.2}$$

In the following we shortly introduce commonly used filters. The *sharp Fourier cutoff filter* (Leonard (1974)) is defined in wave space according to

$$\hat{G}(k) = \begin{cases} 1 & \text{if } k \leq \pi/2 \\ 0 & \text{otherwise} \end{cases}, \tag{A.3}$$

which means that all Fourier modes having a greater wave number k than a specified cutoff are filtered and thus contribute to the unresolved field. The *Gaussian filter*, defined by

$$G(x,x') = \sqrt{\frac{6}{\pi \Delta^2}} \exp\left\{-\frac{6(x-x')^2}{\Delta^2}\right\}, \tag{A.4}$$

where Δ describes the grid size, is often used in combination with a sharp Fourier cutoff filter. It has to be emphasized that the Fourier cutoff filter and the Gaussian filter are invariantly used for homogeneous turbulence and for inhomogeneous turbulence in directions of homogeneity. Since this thesis addresses complex flow geometries, i.e flows which exhibit only one or no homogeneous direction, the aforementioned filter operations are not reasonable to be applied. Therefore the employed unstructured numerical code uses the *box filter*, which reads

$$G(x,x') = \begin{cases} 1/\Delta & \text{if } |x-x'| < \Delta/2 \\ 0 & \text{otherwise} \end{cases}, \tag{A.5}$$

where Δ denotes again the grid spacing.

Appendix B
Optical Devices for Two-Color Laser Induced Fluorescence

The following data have been provided by Laseroptik GmbH, Germany.

B.1 Beamsplitter

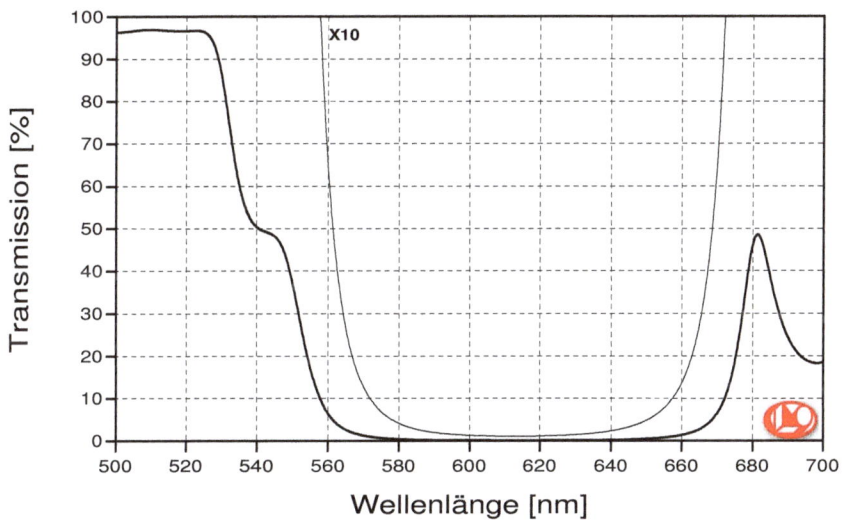

Figure B.1: Optical characteristics of the beamsplitter (arranged at an angle of 45°, with a high reflectance at $\lambda = 575$ nm, and a high transmission at $\lambda = 520$ nm.

B.2 Filter to Remove Mie-Scattering

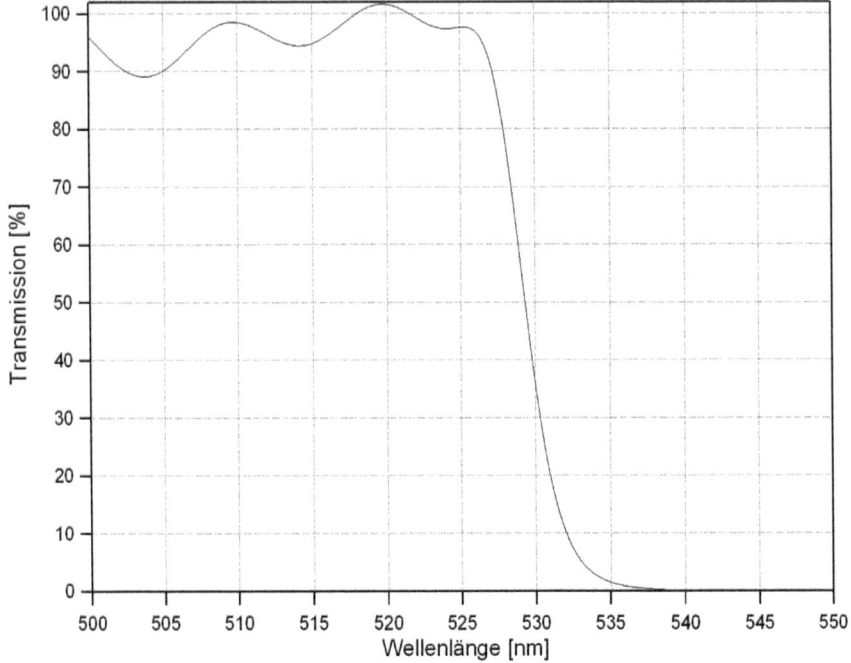

Figure B.2: Optical characteristics of the filter used to remove the Mie-scattering of the seeding particles arranged at an angle of 17°.

Bibliography

Abrams, J. and Hanratty, T. J. (1985). Relaxation effects observed for turbulent flow over a wavy surface, *J. Fluid Mech.* **151**: 443–455.

Adrian, R. J. (1991). Particle–imaging techniques for experimental fluid mechanics, *Annu. Rev. Fluid Mech.* **23**: 261–304.

Adrian, R. J., Christensen, K. T. and Liu, Z.-C. (2000). Analysis and interpretation of instantaneous turbulent velocity fields, *Exp. Fluids* **29**: 275–290.

Alawadhi, E. M. (2005). Forced convection cooling enhancement for rectangular blocks using a wavy plate, *IEEE Transactions on Components and Packaging Technologies* **28**: 525–533.

Angirasa, D., Peterson, G. P. and Pop, I. (1997). Combined heat and mass transfer by natural convection with opposing bouyancy effects in a fluid saturated porous medium, *Int. J. Heat Mass Transfer* **40**(12): 2755–2773.

Ayinde, T. F., Said, S. A. M. and Habib, M. A. (2006). Experimental investigation of turbulent natural convection flow in a channel, *Heat Mass Transfer* **42**(3): 169–177.

Banna, M., Pietri, L. and Zeghmati, B. (2004). Turbulent mixed convection of heat and water vapor transfers in a two-dimensional vegetation canopy, *Heat Mass Transfer* **40**: 757–768.

Belcher, S. E. and Hunt, J. R. C. (1998). Turbulent flow over waves and hills, *Annu. Rev. Fluid Mech.* **30**: 507–538.

Berkooz, G., Holmes, P. and Lumley, J. L. (1993). The proper orthogonal decomposition in the analysis of turbulent flows, *Annu. Rev. Fluid Mech.* **25**: 539–575.

Bojarski, C., Bujko, A., Bujko, R. and Twardowski, R. (1977). Anti-Stokes fluorescence in Rhodamine solutions, *Acta physica et chemica* **23**: 93–99.

Brooke, J. W. and Hanratty, T. J. (1993). Origin of turbulence-producing eddies in a channel flow, *Phys. Fluids* **A5**(4): 1011–1022.

Bruchhausen, M., Guillard, F. and Lemoine, F. (2005). Instantaneous measurement of two-dimensional temperature distributions by means of two-color planar laser induced fluorescence (PLIF), *Exp. Fluids* **38**: 123–131.

Buckles, J., Hanratty, T. J. and Adrian, R. J. (1984). Turbulent flow over large-amplitude wavy surfaces, *J. Fluid Mech.* **140**: 27–44.

Chang, Y. S. and Scotti, A. (2003). Entrainment and suspension of sediments into a turbulent flow over ripples, *J. Turbulence* **4**: 1–22.

Cherukat, P., Na, Y., Hanratty, T. J. and McLaughlin, J. B. (1998). Direct numerical simulation of a fully developed turbulent flow over a wavy wall, *Theoret. Comp. Fluid Dynamics* **11**: 109–134.

Choi, H. S. and Suzuki, K. (2005). Large eddy simulation of turbulent flow and heat transfer in a channel with one wavy wall, *Int. J. Heat Fluid Flow* **26**: 681–694.

Chong, M. S., Perry, A. E. and Cantwell, B. J. (1990). A general classification of three-dimensional flow fields, *Phys. Fluids* **A2**(5): 765–777.

Chu, T. and Goldstein, R. J. (1973). Turbulent natural convection in a horizontal layer of water, *J. Fluid Mech.* **60**: 141–159.

Coolen, M. C. J., Kieft, R. N., Rindt, C. C. M. and van Steenhoven, A. A. (1999). Application of 2D-LIF temperature measurements in water using a Nd:YAG laser, *Exp. Fluids* **27**: 420–426.

Coppeta, J. and Rogers, C. (1998). Dual emission laser induced fluorescence for direct planar scalar behavior measurements, *Exp. Fluids* **25**: 1–15.

Crimaldi, J. P. and Koseff, J. R. (2001). High-resolution measurements of the spatial and temporal scalar structure of a turbulent plume, *Exp. Fluids* **31**: 90–102.

Crimaldi, J. P., Megan, B. W. and Koseff, J. R. (2002). The relationship between mean and instantaneous structure in turbulent passive scalars, *J. Turbulence* **3**: 14.

Dabiri, D. and Gharib, M. (1991). Digital particle image thermometry: The method and implementation, *Exp. Fluids* **11**: 77–86.

Dellil, A. Z., Azzi, A. and Jubran, B. A. (2004). Turbulent flow and convective heat transfer in a wavy wall channel, *Heat Mass Transfer* **40**: 793–799.

Dimotakis, P. E. (2005). Turbulent mixing, *Annu. Rev. Fluid Mech.* **37**: 329–356.

Elsinga, G. E., Scarano, F., Wieneke, B. and van Oudheusden, B. W. (2006). Tomographic particle image velocimetry, *Exp. Fluids* **41**: 933–947.

Evans, G., Greif, R., Siebers, D. and Tieszen, S. (2005). Turbulent mixed convection from a large, high temperature, vertical flat surface, *Int. J. Heat and Fluid Flow* **26**: 1–11.

Fackrell, J. E. and Robins, A. G. (1982). Concentration fluctuations and fluxes in plumes from point sources in a turbulent boundary layer, *J. Fluid Mech.* **117**: 1–26.

Frederick, K. A. and Hanratty, T. J. (1988). Velocity measurements for a turbulent nonseparated flow over solid waves, *Exp. Fluids* **6**: 477–486.

George, W. K. (2007). Is there a universal log law for turbulent wall-bounded flows?, *Phil. Trans. R. Soc.* **365**: 789–806.

Germano, M., Piomeli, U., Moin, P. and Cabot, W. H. (1991). A dynamic subgrid-scale eddy viscosity model, *Phys. Fluids* **A 3**: 1760–1765.

Gnielinski, V. (1976). New equations for heat and mass transfer in turbulent pipe and channel flow, *Int. Chem. Eng.* **16**(2): 359–368.

Günther, A. (2001a). *Large-scale structures in Rayleigh-Bénard convection and flow over waves*, PhD thesis, ETH Zürich, Switzerland.

Günther, A. (2001b). *Large-scale structures in Rayleigh-Bénard convection and flow over waves*, PhD thesis, ETH Zürich, Switzerland.
URL: *http://e-collection.ethbib.ethz.ch/show?type=diss&nr=14359*

Günther, A., Papavassiliou, D. V. and Hanratty, T. J. (1998). Turbulent flow in a channel at low reynolds number, *Exp. Fluids* **25**: 503–511.

Günther, A. and Rudolf von Rohr, P. (2003). Large-scale structures in a developed flow over a wavy wall, *J. Fluid Mech.* **478**: 257–285.

Gong, W., Taylor, P. A. and Dörnbrack, A. (1996). Turbulent boundary-layer flow over fixed aerodynamically rough two-dimensional sinusoidal waves, *J. Fluid Mech.* **312**: 1–37.

Görtler, H. (1940). Über eine dreidimensionale Instabilität laminarer Grenzschichten an konkaven Wänden, *Nachr. Ges. Wiss. Göttingen, Math-Phys. Klasse* **Neue Folge I, 2**: 1–26.

Hanjalić, K. and Kenjereš, S. (2001). 'T-RANS' simulation of deterministic eddy structure in flows driven by thermal buoyancy and Lorentz force, *Flow, Turb. and Comb.* **66**: 427–451.

Hanjalić, K. and Kenjereš, S. (2002). Simulation and identification of deterministic structures in thermal and magnetic convection, *Annals of the New York Academy of Sciences* **972**: 19–28.

Henn, D. S. and Sykes, R. I. (1999). Large-eddy simulation of flow over wavy surfaces, *J. Fluid Mech.* **383**: 75–112.

Hjelmfelt, A. T. and Mockros, L. F. (1966). Motion of discrete particles in a turbulent fluid, *Appl. Sci. Res.* **16**: 149–161.

Hudson, J. D., Dykhno, L. and Hanratty, T. J. (1996). Turbulence production in flow over a wavy wall, *Exp. Fluids* **20**: 257–265.

Hunt, J. C. R. (1985). Turbulent diffusion from sources in complex flows, *Annu. Rev. Fluid Mech.* **17**: 447–485.

Hunt, J. C. R., Puttock, J. S. and Snyder, W. H. (1979). Turbulent diffusion from a point source in stratified and neutral flows around a three-dimensional hill. Part I. Diffusion equation analysis, *Atmos. Environ.* **13**: 1227–1239.

Incropera, F. P. and DeWitt, D. P. (2002). *Fundamentals of heat and mass transfer*, John Wiley & Sons.

Jang, J.-H. and Yan, W.-M. (2004). Mixed convection heat and mass transfer along a vertical wavy surface, *Int. J. Heat Mass Transfer* **47**: 419–428.

Jiménez, J. (2004). Turbulent flows over rough walls, *Annu. Rev. Fluid Mech.* **36**: 173–196.

Karasso, P. S. and Mungal, M. G. (1996). Scalar mixing and reaction in plane liquid shear layers, *J. Fluid Mech.* **323**: 23–63.

Karasso, P. S. and Mungal, M. G. (1997). PLIF measurements in aqueous flows using the Nd:YAG laser, *Exp. Fluids* **23**: 382–387.

Katul, G. G., Finnigan, J. J., Poggi, D., Leuning, R. and Belcher, S. E. (2006). The influence of hilly terrain on canopy-atmosphere carbon dioxide exchange, *Boundary-Layer Meteorology* **118**: 189–216.

Kenjereš, S. and Hanjalić, K. (2006). LES, T-RANS and hybrid simulations of thermal convection at high Ra numbers, *Int. J. Heat Fluid Flow* **27**: 800–810.

Kolmogorov, A. N. (1991). Disspiation of energy in the locally isotropic turbulence, *Phys. Eng. Sci.* **434**: 15–17.

Komori, S., Nagata, K., Kanzaki, T. and Murakami, Y. (1993). Measurements of mass flux in a turbulent liquid flow with a chemical reaction, *AIChE Journal* **39**(10): 1611–1620.

Komori, S., Nagata, K. and Murakami, Y. (1991). Simultaneous measurements of instantaneous concentrations of two reacting species in a turbulent flow with a rapid reaction, *Phys. Fluids* **A3**(4): 507–510.

Krettenauer, K. and Schumann, U. (1989). Direct numerical simulation of thermal convection over a wavy surface, *Meteorol. Atmos. Phys.* **41**: 165–179.

Krogstad, P.-Å. and Antonia, R. A. (1994). Structure of turbulent boundary layers on smooth and rough walls, *J. Fluid Mech.* **277**: 1–21.

Krogstad, P.-Å. and Antonia, R. A. (1999). Surface roughness effects in turbulent boundary layers, *Exp. Fluids* **27**: 450–460.

Krogstad, P.-Å., Antonia, R. A. and Browne, W. B. (1992). Comparison between rough- and smooth-wall turbulent boundary layers, *J. Fluid Mech.* **245**: 599–617.

Kruse, N. (2005). *Isothermal and non-isothermal turbulent flow over solid waves*, PhD thesis, ETH Zurich, Switzerland.
URL: *http://e-collection.ethbib.ethz.ch/show?type=diss&nr=16031*

Kruse, N., Günther, A. and Rudolf von Rohr, P. (2003). Dynamics of large-scale structures in turbulent flow over a wavy wall., *J. Fluid Mech.* **485**: 87–96.

Kruse, N., Kuhn, S. and Rudolf von Rohr, P. (2006). Wavy wall effects on turbulence production and large-scale modes, *J. Turbulence* **7**(31): 1–24.

Kruse, N. and Rudolf von Rohr, P. (2006). Structure of turbulent heat flux in a flow over a heated wavy wall, *Int. J. Heat Mass Transfer* **49**(19-20): 3514–3529.

Kuhn, S., Wagner, C. and Rudolf von Rohr, P. (2007). Influence of wavy surfaces on coherent structures in a turbulent flow, *Exp. Fluids* **43**: 251–259.

Kuzan, J. D., Hanratty, T. J. and Adrian, R. J. (1989). Turbulent flows with incipient separation over solid waves, *Exp. Fluids* **7**: 88–98.

Law, A. W.-K. and Wang, H. (2000). Measurement of mixing processes with combined digital particle image velocimetry and planar laser induced fluorescence, *Exp. Therm. Fluid Sci.* **22**: 213–229.

Lemoine, F., Wolff, M. and Lebouche, M. (1996). Simultaneous concentration and velocity measurements using combined laser-induced fluorescence and laser Doppler velocimetry: Application to turbulent transport, *Exp. Fluids* **20**: 319–327.

Leonard, A. (1974). Energy cascade in large eddy simulation of turbulent fluid flow, *Adv. Geophys.* **18A**: 237–248.

Lin, M. H. and Chen, C. T. (2006). Effect of rotation on the formation of longitudinal vortices in mixed convection flow over a flat plate, *Heat Mass Transfer* **42**(3): 178–186.

Lumley, J. L. and Newman, G. R. (1977). The return to isotropy of homogeneous turbulence, *J. Fluid Mech.* **82**: 161.

Maughan, J. R. and Incropera, F. P. (1989). Regions of heat transfer enhancement for laminar mixed convection in a parallel plate channel, *Int. J. Heat Mass Transfer* **33**: 555–570.

Metwally, H. M. and Manglik, R. M. (2004). Enhanced heat transfer due to curvature-induced lateral vortices in laminar flows in sinusoidal corrugated-plate channels, *Int. J. Heat Mass Transfer* **47**: 2283–2292.

Mitrovic, B. M. and Papavassiliou, D. V. (2004). Effects of a first-order chemical reaction on turbulent mass transfer, *Int. J. Heat Mass Transfer* **47**: 43–61.

Moin, P. and Moser, R. D. (1989). Characteristic-eddy decomposition of turbulence in a channel, *J. Fluid Mech.* **200**: 471–509.

Moulic, S. G. and Yao, L. S. (1989). Mixed convection along a wavy surface, *J. Heat Transfer* **111**: 974–979.

Nagata, K. and Komori, S. (2000). The effects of unstable stratification and mean shear on the chemical reaction in grid turbulence, *J. Fluid Mech.* **408**: 39–52.

Nakagawa, S. and Hanratty, T. J. (2001). Particle image velocimetry measurements of flow over a wavy wall, *Phys. Fluids* **13**(11): 3504–3507.

Nakagawa, S. and Hanratty, T. J. (2003). Influence of a wavy boundary on turbulence. II. Intermediate roughened and hydraulically smooth surface, *Exp. Fluids* **35**: 437–447.

Nakagawa, S., Na, Y. and Hanratty, T. J. (2003). Influence of a wavy boundary on turbulence. I. Highly rough surface, *Exp. Fluids* **35**: 422–436.

Ničeno, B. (2001). *An unstructured parallel algorithm for large eddy and conjugate heat transfer simulations*, PhD thesis, TU Delft.

Ničeno, B. and Nobile, E. (2001). Numerical analysis of fluid flow and heat transfer in periodic wavy channels, *Int. J. Heat Fluid Flow* **22**(2): 156–167.

Osborne, D. G. and Incropera, F. P. (1985a). Experimental study of mixed convection heat transfer for transitional and turbulent flow between horizontal, parallel plates, *Int. J. Heat Mass Transfer* **28**(7): 1337–1344.

Osborne, D. G. and Incropera, F. P. (1985b). Laminar, mixed convection heat transfer for flow between horizontal plates with asymmetric heating, *Int. J. Heat Mass Transfer* **28**: 207–217.

Phillips, W. R. C. (2005). On the spacing of Langmuir circulation in strong shear, *J. Fluid Mech.* **525**: 215–536.

Phillips, W. R. C. and Wu, Z. (1994). On the instability of wave-catalysed longitudinal vortices in strong shear, *J. Fluid Mech.* **272**: 235–254.

Phillips, W. R. C., Wu, Z. and Lumley, J. L. (1996). On the formation of longitudinal vortices in a turbulent boundary layer over wavy terrain, *J. Fluid Mech.* **326**: 321–341.

Poggi, D., Katul, G. G., Albertson, J. D. and Ridolfi, L. (2007). An experimental investigation of turbulent flows over a hilly surface, *Phys. Fluids* **19**: 1–12.

Pope, S. B. (2000). *Turbulent flows*, Cambridge University Press.

Raffel, M., Willert, C. and Kompenhans, J. (1998). *Particle Image Velocimetry. A practical guide*, Springer.

Raupach, M. R., Antonia, R. A. and Rajagopalan, S. (1991). Rough-wall turbulent boundary layers, *Appl. Mech. Rev.* **44**(1): 1–25.

Raupach, M. R. and Finnigan, J. J. (1997). The influence of topography on meteorological variables and surface-atmosphere interactions, *J. Hydrology* **190**: 182–213.

Robinson, S. K. (1991). Coherent motions in the turbulent boundary layer, *Annu. Rev. Fluid Mech.* **23**: 601–639.

Rush, T. A., Newell, T. A. and Jacobi, A. M. (1999). An experimental study of flow and heat transfer in sinusoidal wavy passages, *Int. J. Heat Mass Transfer* **42**: 1541–1553.

Sakakibara, J. and Adrian, R. J. (1999). Whole field measurements of temperature in water using two-color laser-induced fluorescence, *Exp. Fluids* **26**: 7–15.

Sakakibara, J. and Adrian, R. J. (2004). Measurement of temperature field of a Rayleigh-Bénard convection using two-color laser-induced fluorescence, *Exp. Fluids* **37**: 331–340.

Saric, W. S. (1994). Görtler vortices, *Annu. Rev. Fluid Mech.* **26**: 379–409.

Scarano, F. (2002). Iterative image deformation methods in PIV, *Meas. Sci. Technol.* **13**: R1–R19.

Scarano, F. and Riethmuller, M. L. (2000). Advances in iterative multigrid PIV image processing, *Exp. Fluids* **29**: S51–S60.

Schröder, A., Geisler, R., Elsinga, G. E., Scarano, F. and Dierksheide, U. (2008). Investigation of a turbulent spot and tripped turbulent boundary layer flow using time-resolved tomographic PIV, *Exp. Fluids* **in press**.

Shan, J. W., Lang, D. B. and Dimotakis, P. E. (2004). Scalar concentration measurements in liquid-phase flows with pulsed lasers, *Exp. Fluids* **36**: 268–273.

Sirovich, L. (1987). Turbulence and the dynamics of coherent structures. Part I. Coherent structures, *Quart. Appl. Math.* **XLV**: 561–571.

Snyder, W. H. and Hunt, J. C. R. (1984). Turbulent diffusion from a point source in stratified and neutral flows around a three-dimensional hill. Part II. Laboratory measurements of surface concentrations, *Atmos. Environ.* **18**: 1969–2002.

Song, S. and Eaton, J. K. (2004). Reynolds number effects on a turbulent boundary layer with separation, reattachment, and recovery, *Exp. Fluids* **36**: 246–258.

Spalart, P. R. and Allmaras, S. R. (1994). A one-equation turbulence model for aerodynamic flows, *La Recherche Aérospatiale* **1**: 5–21.

Stalio, E. and Nobile, E. (2003). Direct numerical simulation of heat transfer over ribblets, *Int. J. Heat Fluid Flow* **24**: 356–371.

Su, L. K. and Mungal, M. G. (2004). Simultaneous measurements of scalar and velocity field evolution in turbulent crossflowing jets, *J. Fluid Mech.* **513**: 1–45.

Thorsness, C. B., Morrisroe, P. E. and Hanratty, T. J. (1978). A comparison of linear theory with measurements of the variation of shear stress along a solid wave, *Chem. Eng. Sci.* **33**: 579–592.

Tseng, Y.-H. and Ferziger, J. H. (2004). Large-eddy simulation of turbulent wavy boundary flow - illustration of vortex dynamics, *J. Turbulence* **5**(034): 1–23.

Wagner, C. (2007). *Transport Phenomena in complex turbulent flows: Numerical and experimental methods*, PhD thesis, ETH Zurich, Switzerland.

Wagner, C., Kuhn, S. and Rudolf von Rohr, P. (2007). Scalar transport from a point source in flows over wavy walls, *Exp. Fluids* **43**: 261–271.

Wagner, W. and Kruse, A. (1998). *Properties of water and steam. The industrial standard IAPWS-IF97 for thermodynamic properties and supplementary equations for other properties*, Springer Verlag Berlin.

Warhaft, Z. (2000). Passive scalars in turbulent flows, *Annu. Rev. Fluid Mech.* **32**: 203–240.

Westerweel, J. (1993). *Digital particle image velocimetry - Theorie and application -*, PhD thesis, TU Delft.

Westerweel, J. (1997). Fundamentals of digital particle image velocimetry, *Meas. Sci. Technol.* **8**: 1379–1392.

Westerweel, J. and van Oord, J. (2000). Stereoscopic PIV measurements in a turbulent boundary layer, *in* M. Stanislas, J. Kompenhans and J. Westerweel (eds), *Particle image velocimetry, progress toward industrial application*, Kluwer, Dortrecht.

Yao, L. S. (1983). Natural convection along a vertical wavy surface, *J. Heat Transfer* **105**: 465–468.

Yu, C. H., Chang, M. Y., Huang, C. C. and Lin, T. F. (1997). Unsteady vortex roll structures in a mixed convective air flow through a horizontal plane channel: a numerical study, *Int. J. Heat Mass Transfer* **40**(3): 505–518.

Yu, C. H., Chang, M. Y. and Lin, T. F. (1997). Structures of moving transverse and mixed rolls in mixed convection of air in a horizontal plane channel, *Int. J. Heat Mass Transfer* **40**(2): 333–346.

Zedler, E. A. and Street, R. L. (2001). Large-eddy simulation of sediment transport: Currents over ripples, *J. Hydraulic Eng.* **127**: 444–452.

Zhang, H., Huang, X. Y., Li, H. S. and Chua, L. P. (2002). Flow patterns and heat transfer enhancement in low-Reynolds-Rayleigh-number channel flow, *Appl. Therm. Eng.* **22**: 1277–1288.

Zhou, J., Adrian, R. J., Balachandar, S. and Kendall, T. M. (1999). Mechanism for generating coherent packets of hairpin vortices in channel flow, *J. Fluid Mech.* **387**: 353–396.

Zilker, D. P., Cook, G. W. and Hanratty, T. J. (1977). Influence of the amplitude of a solid wavy wall on a turbulent flow. Part 2. Non-separated flows, *J. Fluid Mech.* **82**(part 1): 29–51.

Zilker, D. P. and Hanratty, T. J. (1979). Influence of the amplitude of a solid wavy wall on a turbulent flow. Part 1. Separated flows, *J. Fluid Mech.* **90**(part 2): 257–271.

Acknowledgements

This research project was conducted at the Institute of Process Engineering at ETH Zurich. The project was funded with financial support from the Swiss National Science Foundation (SNF project 109560). The numerical work was carried out at TU Delft under the HPC-EUROPA project (RII3-CT-2003-506079), with the support of the European Community - Research Infrastructure Action under the FP6 'Structuring the European Research Area' Programme.

I wish to express my gratitude to my advisor, Prof. Philipp Rudolf von Rohr, for his support during this thesis, for the opportunity to work on this fundamental study, and for all the freedom in research he provided. I thank Prof. Thomas Rösgen and Dr. Saša Kenjereš for accepting the task as co-advisors of this thesis, and for their interest and their contributions to this project. In addition, I want to thank Saša for the nice time sharing his office in Delft and for sharing his immense expertise in numerical computations.

My interest in fluid dynamics was raised during my course of studies at TU Munich. In this context I wish to thank Prof. Johann Stichlmair who introduced to me the fascination and relevance of fluid dynamics in technical applications, and for all his personal support which allowed me to spend two semesters at MIT and at ETH Zurich.

I want to thank my predecessor Dr. Nils Kruse for passing all of his knowledge to me which allowed me a quick-start into this project, for the many interesting discussions, and for the interest he still has in this project. Also many thanks to the colleagues who shared some time with me in Schlieren, Dr. Carsten Wagner, Dr. Severin Wälchli and Dr. Adrian Wegmann. I also would like to thank all former and present members of the Transport Processes and Reactions Laboratory for many discussions and for creating such a nice atmosphere: Dr. Rodrigo Amandi, Dr. Cordin Arpagaus, Dr. Andrea Bieder, Dr. Beat Borer, Dr. Serge Deportes, Dr. Andrea Grüniger, Dr. Karol Prikopsky, Dr. Axel Sonnenfeld, Dr. Michael Studer, Dr. Beat Wellig, Donata Maria Fries, Cédric Hutter, Lutz Körner, Patrick Reichen, Tobias Rothenfluh, Martin Schuler, Adrian Spillmann, Bruno Tidona, Franz Trachsel, Tobias Voitl. In addition, I thank all the students who contributed to this work with a semester or diploma thesis: Michael Meier, Jeannot Maier, Tobias Rothenfluh and Adrian Zenklusen.

I thank Silvia Christoffel for taking care of all the administrative issues, and the people of the workshop of the institute, Peter Hoffmann, Bruno Kramer, René Plüss and Christian Rohrbach, for many discussions and their help in constructing and maintaining the flow facility. Especially many thanks to Bruno, who spend hours in milling the wavy surfaces to meet our special needs, and for assembling that large number of channel sections used in this thesis. I thank Sascha Jovanovic and Beni Cadonau for solving any IT issue.

I thank my family for their help, support and belief, which made my education and finally this thesis possible.

And finally and most importantly, I wish to express my warmest thanks to Sarah-Maria for her continuous support, patience and love.

VDM Verlagsservicegesellschaft mbH

Die VDM Verlagsservicegesellschaft sucht für wissenschaftliche Verlage abgeschlossene und herausragende

Dissertationen, Habilitationen, Diplomarbeiten, Master Theses, Magisterarbeiten usw.

für die kostenlose Publikation als Fachbuch.

Sie verfügen über eine Arbeit, die hohen inhaltlichen und formalen Ansprüchen genügt, und haben Interesse an einer honorarvergüteten Publikation?

Dann senden Sie bitte erste Informationen über sich und Ihre Arbeit per Email an *info@vdm-vsg.de*.

Sie erhalten kurzfristig unser Feedback!

VDM Verlagsservicegesellschaft mbH
Dudweiler Landstr. 99
D - 66123 Saarbrücken

Telefon +49 681 3720 174
Fax +49 681 3720 1749

www.vdm-vsg.de

Die VDM Verlagsservicegesellschaft mbH vertritt

Printed by Books on Demand GmbH, Norderstedt / Germany